Mathematics

The HighScope Preschool Curriculum

Mathematics

Ann S. Epstein, PhD

HIGHSCOPE
PRESS ®

Ypsilanti, Michigan

Published by
HighScope® Press

A division of the
HighScope Educational Research Foundation
600 North River Street
Ypsilanti, Michigan 48198-2898
734.485.2000, FAX 734.485.0704
Orders: 800.40.PRESS; Fax: 800.442.4FAX; www.highscope.org
E-mail: *press@highscope.org*

Editor: Marcella Fecteau Weiner

`

Photography:
Bob Foran — Front cover, 1, 7, 9 (bottom), 17, 24 (KDI 33, KDI 35), 25 (KDI 37), 29, 30, 32, 35, 39, 45 (bottom right), 47 (right), 53, 59, 60, 61, 63, 67, 72, 78, 82, 85, 89, 91, 92, 95, 98 (top right), 103, 104, 106, 107, 109, 110, 113, 114, back cover
Gregory Fox — 5, 13, 15, 19, 22, 24 (KDI 31, KDI 32), 25 (KDI 39), 26, 36, 40, 42, 45 (top, bottom left), 65, 71, 73, 75, 81, 98 (bottom), 100, 116, 119
Pat Thompson — 123
HighScope Staff — All other photos

Library of Congress Cataloging-in-Publication Data
Epstein, Ann S.
  Mathematics / Ann S. Epstein, PhD.
    pages cm. -- (The HighScope preschool curriculum)
  Includes bibliographical references.
  ISBN 978-1-57379-655-2 (soft cover : alk. paper) 1. Mathematics--Study and teaching (Preschool) 2. Mathematics--Curricula. I. Title.
  QA135.6.E576 2012
  372.7'044--dc23
                                    2012007722

Printed in the United States of America
10 9 8 7 6 5 4 3 2 1

Contents

Acknowledgments

Many people contributed their knowledge and skills to the publication of *Mathematics*. I want to thank the Early Childhood staff as well as other HighScope staff members who collaborated on creating the key developmental indicators (KDIs) in this content area: Beth Marshall, Sue Gainsley, Shannon Lockhart, Polly Neill, Kay Rush, Julie Hoelscher, and Emily Thompson. Among this group of colleagues, those who devoted special attention to reviewing the manuscript for this book were Emily Thompson and Beth Marshall. Mary Hohmann, whose expertise informs many other books about the HighScope Curriculum, also provided detailed feedback.

The developmental scaffolding charts in this volume — describing what children might do and say and how adults can support and gently extend their learning at different developmental levels — are invaluable contributions to the curriculum. I am grateful to Beth Marshall and Sue Gainsley for the extraordinary working relationship we forged in creating these charts. By bringing our unique experiences to this challenging process, we integrated knowledge about child development and effective classroom practices from the perspectives of research, teaching, training, and policy.

Thanks are also due to Nancy Brickman, who directed the editing and production of the book. I extend particular appreciation to Marcella Fecteau Weiner, who edited the volume, and Katie Bruckner, who assisted with all aspects of the publication process. I also want to acknowledge the following individuals for contributing to the book's visual appeal and reader friendliness: photographers Bob Foran and Gregory Fox, and graphic artists Judy Seling (book design) and Kazuko Sacks (book production).

Finally, I extend sincerest thanks to all the teachers, trainers, children, and families whose participation in HighScope and other early childhood programs has contributed to the creation and authenticity of the HighScope Preschool Curriculum over the decades. I hope this book continues to support children's learning in mathematics for many years to come.

CHAPTER 1

The Importance of Mathematics

Early mathematics learning is currently receiving the attention once lavished on literacy. Not only are educators developing more respect for what young children are capable of in this area, but they are also realizing that the thinking and reasoning that underlie mathematics are critical for learning in other areas as well. For example, reading and writing numbers involves literacy, spatial awareness is a component of physical coordination, timing turns or dividing something to share is social problem solving, gathering and analyzing data are the bedrock of science, and cutting materials to size and fitting them together are useful in making props for pretend play.

There is also a growing awareness, supported by research, that mathematics in the early years is much more than numbers and rote counting. Exploring shapes is the beginning of geometry, asking "how many?" is measurement, exploring patterns is algebra, and collecting information to answer everyday questions ("Who likes what in their trail mix?") is practical data analysis.

Children enjoy and are eager to learn mathematics. In fact, they are often more confident and adventurous than many adults, for whom "math anxiety" was an unfortunate part of growing up. When we as adults recognize how much we use mathematics in our own daily activities (paying bills, calculating mileage, planning a garden, building a toolshed), we can more easily make appropriate materials and hands-on activities available to young children. Then, as illustrated by the anecdotes in "Mathematics in Action" (see box on this page), Zeke can count the eyes in his robot, Patrick can find shapes in his snacks, Kendra can make patterns, Natreal can chant position words while she swings, Adam can measure the length of a worm, and Isobel can find a toy couch that is just the right size.

Mathematics in Action

At small-group time, Zeke counts the seven marbles he inserts as eyes in his play dough robot.

❖

At snacktime, while rotating a cheese cracker, Patrick says, "Hey it's a diamond. Now it's a square!"

❖

At small-group time Kendra plays with the Rig-A-Ma-Jigs. "I made a pattern!" she says excitedly. "Red, yellow, red, yellow, red, yellow."

❖

At outside time, Natreal repeats, "Backward, forward, backward, forward" as she swings.

❖

At small-group time, Maeve (a teacher) holds a plastic worm while Adam uses a tape measure. When Maeve asks how long the worm is, Adam says, "It's four." (The worm is four inches.)

❖

At work time in the toy area, Isobel puts a small piece of cloth on the doll couch and says, "This goes on a smaller couch." She hunts in the basket of doll furniture until she finds a smaller couch. "This fits perfectly," she says, putting the small cloth on the small couch.

An Emerging Appreciation for Early Mathematics Learning

The importance of mathematics in early education is only now getting the attention it deserves. According to the National Research Council's Committee on Early Childhood Mathematics (2009), "mathematics has risen to the top of the national policy agenda as part of the need to improve the technical and scientific literacy of the American public…. There is particular concern about the chronically low mathematics and science performance of economically disadvantaged students and the lack of diversity in the science and technical workforce. Particularly alarming is that such disparities exist in the earliest years of schooling and even before school entry" (p. 1).

Although policymakers are concerned about the low mathematical performance of young children, particularly disadvantaged children, this low performance cannot be connected to young children's lack of interest in mathematics. In fact, in the past 25 years, researchers have come to recognize and appreciate how much young children enjoy and are capable of doing mathematics. For example, when Herbert Ginsburg and his colleagues observed children's free play, they were amazed not only by how much the children used mathematical ideas but also by how advanced their thinking was (Ginsburg, Inoue, & Seo, 1999). Professor Arthur Baroody (2000) found that preschoolers actively built mathematical knowledge from their daily experiences and then applied the concepts they derived to solving the problems they encountered in play and work. Researchers have accumulated a wealth of evidence that children between the ages of three and five years of age construct a variety of fundamentally important

> "Mathematical experiences for very young children should build largely upon their play and the natural relationships between learning and life in their daily activities, interests, and questions."
> — Clements (2004b, p. 59)

informal mathematical concepts and strategies from their everyday experiences. Indeed, this evidence indicates that they are predisposed, perhaps innately, to attend to numerical situations and problems (Baroody, 2000).

As noted previously, early mathematics involves more than reciting numerals or rote counting, although these limited options are often the only ones provided in many preschool programs. Observations of young children's spontaneous mathematical activities show that investigating numbers accounts for only part of their curiosity. Preschoolers are at least, if not more, interested in manipulating shapes, discovering and describing patterns, and exploring the transformations brought about by processes like adding and subtracting (Ginsburg et al., 1999).

In fact, young children, like adults, use mathematics every day, often without realizing it. Simple, age-appropriate activities, such as building with blocks or asking who is older or taller, set the stage for later learning in geometry or measurement, respectively. Preschoolers are beginning to understand what numbers are and how they work, not by saying their 1, 2, 3s, but by counting real things, such as the number of tomatoes growing in the garden. When toddlers empty and refill containers, or preschoolers build a road with blocks, they are laying the foundation for geometric concepts such as volume and area. Figuring out who jumped farther involves comparing distances, a precursor to measurement. Looking for patterns in a design or a series of movements prepares preschoolers for numerical relations involved in middle school algebra. And finding out who wants more pretzels or more raisins in their trail mix is applying data analysis to real-life decisions.

Because we use mathematics in our everyday lives — to pay bills, follow a recipe, calculate the supplies needed for a construction project, or read a map — developing mathematical competence is an essential goal of education. Positive and effective early mathematics experiences not only help children develop specific computational and analytical skills but also engender a healthy and welcoming attitude toward this core subject. In addition, linking mathematics to other domains (what the National Council of Teachers of Mathematics [NCTM] calls "making connections") helps to foster learning across content areas. For example, reading and writing numbers involves both literacy and mathematics. Choosing the colors and arrangement for a patterned border is also an artistic exercise. Deciding to use a sand timer to take turns on the computer brings mathematics into social problem solving. Given these connections, it is therefore not surprising that preschool children's mathematical abilities are an important predictor of their later school success in all areas of the curriculum (Duncan et al., 2007).

Math Is Everywhere

With young children, opportunities to investigate mathematics are everywhere. Math learning is taking place whenever the following activities are going on:

- **Observing:** Discovering and creating knowledge about the world using all the senses; for example, seeing colors and observing some are lighter or darker than others, hearing sounds and noticing that they vary in pitch and loudness

- **Exploring materials:** Discovering the properties of objects and how things work and seeing how things change when they are acted upon by people or events; for example, seeing what happens when a blob of blue paint is mixed with a blob of yellow paint, watching the computer screen display two dogs after touching the numeral 2

- **Working with numbers:** Making intuitive judgments about quantity without counting, understanding that numerals represent numbers of objects, grasping one-to-one correspondence, and counting

- **Ordering things:** Putting things in order according to some graduated attribute on which they differ such as size, age, loudness, color intensity, and so on; for example, hanging pots on a pegboard from biggest to smallest

- **Navigating in space:** Arranging objects, understanding how one's body relates to its surroundings, fitting things together and taking them apart, and understanding direction and position concepts; for example, putting the widest block at the base of a tower, doing a puzzle, building a hideout under the table

- **Comparing quantities:** Recognizing bigger, smaller; more, fewer/less; and so on; comparing amounts in continuous materials (such as sand and water) and discrete objects (such as blocks and beads); for example, piling up sand to make big and little sand cakes, counting snaps to see if their shirt or a friend's has more snaps down the front

- **Identifying regularities:** Recognizing, copying, adding to, and creating patterns; identifying regularity and repetition in objects and events; making predictions based on observed patterns; for example, gluing a row of alternating red and blue circles, knowing that outside time always comes after snacktime in the daily routine

Just like adults, young children use mathematics every day without even realizing it. This child is setting the table for mealtime, which involves counting and one-to-one correspondence.

> "The challenge in early mathematics education is to use children's initial understandings to proactively support their *continued* mathematical development."
>
> — Campbell (1997, p. 106)

and, in fact, to accomplishing most tasks.

Educator and researcher Juanita V. Copley (2010) notes that "young children are typically motivated to learn quantitative and spatial information. Their dispositions allow them to be positive and confident in their mathematical abilities. The highly prized characteristics of persistence, focused participation, hypothesis testing, risk taking, and self-regulation are often present, but seldom acknowledged, in the young child" (p. 7). Creating a positive mathematics environment in the preschool classroom can therefore contribute to children's specific quantitative abilities, as well as promoting a problem-solving orientation to learning in general.

Likewise, other early abilities, such as *executive functioning* (i.e., planning and decision making), during preschool predict later mathematical achievement in the elementary grades (Clark, Pritchard, & Woodward, 2010).

Moreover, early mathematics education is important for the broader critical thinking skills it develops in young children, which they can then apply to virtually every other area of learning. Mathematical processes such as calculating, estimating, predicting, transforming, and making comparisons are also relevant to comprehending the arc of a story, investigating nature, moving one's body through space or to music, or figuring how many jumps it will take to cross the yard. The idea that a problem can be solved in more than one way is not only a mathematical insight, it is also vital to solving social conflicts

Components of Mathematics

The National Council of Teachers of Mathematics (2000) identifies five content areas for mathematics learning across all age groups: number and operations, geometry, measurement, algebra, and data analysis. The first three areas, called *focal points,* are especially important in the early childhood years (National Council of Teachers of Mathematics, 2006). The in-depth Numbers Plus Preschool Mathematics Curriculum developed by HighScope (Epstein, 2009) is based on the NCTM's content areas and focal points, as are the mathematics key developmental indicators (KDIs) described later in this chapter and detailed in this book.

Mathematics Content Areas

- **Number and operations** involves understanding whole numbers and realizing that numbers represent quantity. It includes learning number words and symbols, counting, comparing and ordering quantities, composing (combining) and decomposing (dividing) numbers, and simple addition and subtraction.

- **Geometry** involves identifying shapes and describing spatial relationships and includes learning the names and properties of two- and three-dimensional shapes, transformation (changing shapes, putting them together and taking them apart), and spatial reasoning (using position, direction, and distance words).

- **Measurement** involves identifying measurable attributes and using them to compare objects, actions, people, and events. For young children, this means learning simple measurement terms and processes, understanding what a *unit* is, and comparing and ordering attributes.

- **Algebra** involves identifying patterns and relationships, including describing, copying, and creating simple alternating patterns (such as ABABAB, ABCABCABC, or AABAABAAB) and recognizing and describing increasing and decreasing patterns (such as cycles of plant growth or children getting older).

- **Data analysis** involves formulating and answering questions by collecting, organizing, and analyzing information. In the early years, this includes describing attributes, organizing and comparing simple data, representing findings on simple charts or graphs, and interpreting and applying the lessons learned.

Early geometry includes learning the names and properties of different shapes.

In addition to having a better grasp of *what* young children understand, educators also know more about *how* to foster early mathematics learning (National Mathematics Advisory Panel, 2008). Opportunities to build on children's interest in mathematics abound throughout the program environment and daily routine. However, just because "math is everywhere," it does not mean learning can be haphazard or left to chance. In fact, it is often the case that little of the young child's mathematical explorations register "because teachers lack developmental knowledge of children's math learning" (Copple, 2004, p. 85). Adults must systematically engage preschoolers in working with materials, pursuing investigations, and using mathematical thinking to draw conclusions. Teachers can encourage mathematics reasoning with thoughtful comments such as

"I wonder what would happen if…" and "Why do you think…?" (See the teaching strategies described in chapter 2.) This view is highlighted in *Early Childhood Mathematics: Promoting Good Beginnings*, the joint position paper of the National Association for the Education of Young Children (NAEYC) and NCTM (2002): "Because young children's experiences fundamentally shape their attitude toward mathematics, an engaging and encouraging climate for children's early encounters with mathematics is important. It is vital for young children to develop confidence in their ability to understand and use mathematics — in other words, to see mathematics as within their reach" (p. 4).

One important aspect of teaching mathematics is understanding that children learn sequentially (Campbell, 1997; Clements, Sarama, & DiBiase, 2004); that is, each new concept

Adults encourage children's mathematical thinking with thoughtful comments and questions, such as "Why do you think that puzzle piece goes there?"

or skill builds on what children have learned before. Most mathematics curricula have a fixed sequence of activities that all the children in the class proceed through together. An alternative approach (the one used in Numbers Plus) offers open-ended activities that allow children to engage with the materials and ideas at their own level in the learning sequence. For a broader discussion of the benefits of sequencing activities within rather than across activities, see chapter 9 in *The HighScope Preschool Curriculum* (Epstein & Hohmann, 2012).

Mathematics is part of our daily lives. By understanding how young children engage with this content area, and how we as adults can overcome our reservations to share in their excitement about it, we can establish a foundation for a lifetime of mathematical competence.

About This Book

In the HighScope Preschool Curriculum, the content of children's learning is organized into eight areas: A. Approaches to Learning; B. Social and Emotional Development; C. Physical Development and Health; D. Language, Literacy, and Communication; E. Mathematics; F. Creative Arts; G. Science and Technology; and H. Social Studies. Within each content area, HighScope identifies **key developmental indicators (KDIs)** that are the building blocks of young children's thinking and reasoning.

The term *key developmental indicators* encapsulates HighScope's approach to early education. The word *key* refers to the fact that these are the meaningful ideas children should learn and experience. The second part of the term — *developmental* — conveys the idea that learning is gradual and cumulative. Learning follows a sequence, generally moving from simple to more complex knowledge and skills. Finally, we chose the term *indicators* to emphasize that educators need evidence that children are developing the knowledge, skills, and understanding considered important for school and life readiness. To plan appropriately for students and to evaluate program effectiveness, we need observable indicators of our impact on children.

This book is designed to help you as you guide and support young children's learning in the Mathematics content area in the HighScope Curriculum. This chapter provided insights from research literature on how children approach learning and summarized basic principles of how children acquire knowledge and skills. Chapter 2 describes general teaching strategies for Mathematics and provides an overview of the KDIs for this content area.

Chapters 3–11, respectively, provide teaching strategies for each of the nine KDIs in Mathematics:

31. **Number words and symbols:** Children recognize and use number words and symbols.

32. **Counting:** Children count things.

33. **Part-whole relationships:** Children combine and separate quantities of objects.

34. **Shapes:** Children identify, name, and describe shapes.

35. **Spatial awareness:** Children recognize spatial relationships among people and objects.

36. **Measuring:** Children measure to describe, compare, and order things.

37. **Unit:** Children understand and use the concept of unit.

38. **Patterns:** Children identify, describe, copy, complete, and create patterns.

39. **Data analysis:** Children use information about quantity to draw conclusions, make decisions, and solve problems.

At the end of each of these chapters is a chart showing ideas for scaffolding learning for that KDI. The chart will help you recognize the specific abilities that are developing at earlier, middle, and later stages of development and gives corresponding teaching strategies that you can use to support and gently extend children's learning at each stage.

In HighScope programs, teachers support children in open-ended activities in all the content areas (including Mathematics) so that children can engage with the materials at their own level.

HighScope Preschool Curriculum Content
Key Developmental Indicators

A. Approaches to Learning

1. **Initiative:** Children demonstrate initiative as they explore their world.

2. **Planning:** Children make plans and follow through on their intentions.

3. **Engagement:** Children focus on activities that interest them.

4. **Problem solving:** Children solve problems encountered in play.

5. **Use of resources:** Children gather information and formulate ideas about their world.

6. **Reflection:** Children reflect on their experiences.

B. Social and Emotional Development

7. **Self-identity:** Children have a positive self-identity.

8. **Sense of competence:** Children feel they are competent.

9. **Emotions:** Children recognize, label, and regulate their feelings.

10. **Empathy:** Children demonstrate empathy toward others.

11. **Community:** Children participate in the community of the classroom.

12. **Building relationships:** Children build relationships with other children and adults.

13. **Cooperative play:** Children engage in cooperative play.

14. **Moral development:** Children develop an internal sense of right and wrong.

15. **Conflict resolution:** Children resolve social conflicts.

C. Physical Development and Health

16. **Gross-motor skills:** Children demonstrate strength, flexibility, balance, and timing in using their large muscles.

17. **Fine-motor skills:** Children demonstrate dexterity and hand-eye coordination in using their small muscles.

18. **Body awareness:** Children know about their bodies and how to navigate them in space.

19. **Personal care:** Children carry out personal care routines on their own.

20. **Healthy behavior:** Children engage in healthy practices.

D. Language, Literacy, and Communication[1]

21. **Comprehension:** Children understand language.

22. **Speaking:** Children express themselves using language.

23. **Vocabulary:** Children understand and use a variety of words and phrases.

24. **Phonological awareness:** Children identify distinct sounds in spoken language.

25. **Alphabetic knowledge:** Children identify letter names and their sounds.

26. **Reading:** Children read for pleasure and information.

27. **Concepts about print:** Children demonstrate knowledge about environmental print.

28. **Book knowledge:** Children demonstrate knowledge about books.

29. **Writing:** Children write for many different purposes.

30. **English language learning:** (If applicable) Children use English and their home language(s) (including sign language).

---

[1]Language, Literacy, and Communication KDIs 21–29 may be used for the child's home language(s) as well as English. KDI 30 refers specifically to English language learning.

E. Mathematics

31. **Number words and symbols**: Children recognize and use number words and symbols.

32. **Counting**: Children count things.

33. **Part-whole relationships**: Children combine and separate quantities of objects.

34. **Shapes**: Children identify, name, and describe shapes.

35. **Spatial awareness**: Children recognize spatial relationships among people and objects.

36. **Measuring**: Children measure to describe, compare, and order things.

37. **Unit**: Children understand and use the concept of unit.

38. **Patterns**: Children identify, describe, copy, complete, and create patterns.

39. **Data analysis**: Children use information about quantity to draw conclusions, make decisions, and solve problems.

F. Creative Arts

40. **Art**: Children express and represent what they observe, think, imagine, and feel through two- and three-dimensional art.

41. **Music**: Children express and represent what they observe, think, imagine, and feel through music.

42. **Movement**: Children express and represent what they observe, think, imagine, and feel through movement.

43. **Pretend play**: Children express and represent what they observe, think, imagine, and feel through pretend play.

44. **Appreciating the arts**: Children appreciate the creative arts.

G. Science and Technology

45. **Observing**: Children observe the materials and processes in their environment.

46. **Classifying**: Children classify materials, actions, people, and events.

47. **Experimenting**: Children experiment to test their ideas.

48. **Predicting**: Children predict what they expect will happen.

49. **Drawing conclusions**: Children draw conclusions based on their experiences and observations.

50. **Communicating ideas**: Children communicate their ideas about the characteristics of things and how they work.

51. **Natural and physical world**: Children gather knowledge about the natural and physical world.

52. **Tools and technology**: Children explore and use tools and technology.

H. Social Studies

53. **Diversity**: Children understand that people have diverse characteristics, interests, and abilities.

54. **Community roles**: Children recognize that people have different roles and functions in the community.

55. **Decision making**: Children participate in making classroom decisions.

56. **Geography**: Children recognize and interpret features and locations in their environment.

57. **History**: Children understand past, present, and future.

58. **Ecology**: Children understand the importance of taking care of their environment.

CHAPTER 2

General Teaching Strategies
for Mathematics

Young children develop mathematical knowledge and skills through meaningful interactions with materials, people, events, and ideas. However, just because "math is everywhere," it does not mean that children will automatically acquire mathematical understanding. Teachers must be active in providing appropriate materials and encouraging reflective thinking to help preschoolers construct mathematical knowledge from their experiences. To be intentional in supporting early mathematics learning, use the following overall strategies listed here as well as the specific ideas provided in each key developmental indicator (KDI) chapter.

General Teaching Strategies

Provide a wide variety of mathematics materials in every area of the classroom

You do not need to create a separate mathematics area in the classroom. Instead, create an environment that ensures children will "bump into interesting mathematics at every turn" (Greenes, 1999, p. 46). To help turn "mathematics is everywhere" into a reality, provide preschoolers with a variety of quantitative materials (things that can be counted or measured), such as

- Kitchen timers, spinners, and number books (for number words and symbols)

- Small countable toys and pegs with pegboards (for counting)

- Collections of small items such as blocks or shells that children can combine and separate (for part-whole relationships)

- Two- and three-dimensional shapes children can handle (for shapes)

- Large structures children can climb over, under, around, and through (for spatial awareness)

- Measuring cups and spoons (for measuring)

- Lengths of string and rulers (for unconventional and conventional measuring units)

- Beads and noisemakers (to create visual and sound patterns)

- Paper and markers (to record and analyze data)

(See "Recommended Materials for Mathematics" on pp. 16–17 for other materials that support the Mathematics KDIs.) The following anecdotes illustrate how preschoolers use these quantitative materials in their everyday activities:

At work time in the block area, as Delaney drops cars down a tube, she counts them from 1 to 19.

❖

At small-group time, Chan makes shapes like the ones in Mouse Shapes *by Ellen Stoll Walsh. He makes a square with two rectangular blocks.*

❖

At work time in the art area, Cindy looks for a circle to make the snowman's base. Joe gets a wooden one from the toy area and says, "I found you a big circle, Cindy."

❖

At work time in the house area, Joey identifies the numerals 1 through 8 on the computer keyboard. He asks, "What's this?" as he points to the 9.

❖

At snacktime, Wendy points to a pretzel and says, "That's the number 8. It's lying down."

❖

At large-group time, Madra holds a scarf in each hand and says, "Now I have two."

Computers and other forms of technology can play a useful but limited role in early mathematics education (as well as science and other content areas), if they are used appropriately. Software should be open ended and promote discovery rather than rote drill and practice. Programs can be especially effective when adults work alongside children and help them think through the process of how they arrive at an answer. Children can also work together at the computer. According to researcher Douglas Clements (1999), "contrary to initial fears, computers do not isolate children. Rather they serve as potential catalysts for social interaction" (p. 122). When children share computers, they collaborate to solve problems, talk about what

they are doing, help and teach friends, and create rules for cooperation. In fact, they prefer working on the computer with a friend to doing so alone. For more information on the appropriate use of technology with preschoolers, see the companion book, *Science and Technology* (Epstein, 2012b).

Finally, be sure to give children sufficient time and space to discover the mathematical properties of whatever materials you provide, even those you never intended for this purpose because "the ways children use objects are often very different than those we intend or define" (Seo, 2003, p. 30). For example, preschoolers may put the lids of markers on their fingertips to explore one-to-one correspondence

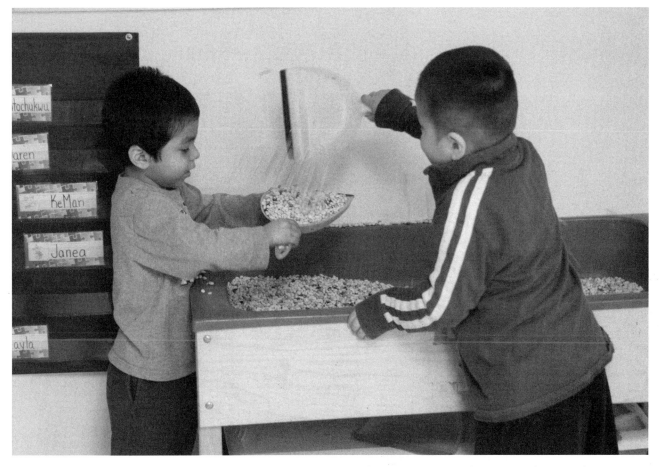

Let children explore and discover the mathematical properties of whatever materials you provide, even those provided for a different purpose. Here these two boys use dust pans (typically used for cleaning up) as a measuring tool for the beans in the sand and water table.

Recommended Materials for Mathematics

Materials for number sense and operations (KDIs 31, 32, and 33)

- Printed materials containing numerals and other mathematical symbols, such as signs, labels, brochures, and advertisements

- Things with numerals on them, such as calculators, playing cards, thermometers, and simple board games with dice or spinners

- Numerals made of wood, plastic, or heavy cardboard (so children can sort, copy, and trace them)

- Discrete items children can count, such as beads, blocks, shells, cubes, bottle caps, and silverware

- Materials to create one-to-one correspondence, such as pegs and pegboards, colored markers and tops, nuts and bolts, and cups and saucers

- Books about numbers and creating different-sized sets by adding and subtracting

Materials for geometry (KDIs 34 and 35)

- Materials and tools for filling and emptying, such as water, sand, scoops, and shovels

- Commercial and household materials to fit together and take apart, such as Legos, Tinkertoys, puzzles, felt pieces, boxes and lids, and clothing with different types of fasteners

- Attribute blocks that vary in shape, size, color, and thickness

- Two- and three-dimensional geometric shapes in a variety of sizes, such as circles, triangles, and rectangles (including squares), spheres, cubes, and cones; triangles should include right triangles (one angle equal to 90 degrees), acute triangles (one angle less than 90 degrees), isosceles triangles (two equal angles and sides), equilateral triangles (three equal angles and sides), and obtuse triangles (one angle greater than 90 degrees)

- Tangram pieces

- Wooden and sturdy cardboard blocks in conventional and unconventional shapes

- Containers and covers in different shapes and sizes (recycled containers from home work well)

- Materials to create two-dimensional shapes, such as string, pipe cleaners, and yarn

- Moldable materials to create three-dimensional shapes, such as clay, dough, sand, and beeswax

- Things with moving parts, such as kitchen utensils, musical instruments, and cameras

- Books that feature shapes and locations with illustrations from different perspectives

- Photos of classroom materials and activities from different viewpoints

- Materials that change with manipulation or time, such as clay, play dough, computer drawing programs, sand, water, plants, and animals

- Materials to explore spatial concepts (e.g., in, out, over, under, up, down, around, through) and to view things from different heights and positions, such as climbing equipment, large empty cartons (from appliances or furniture), and boards

- Materials with vertical and horizontal symmetry, such as doll clothes, a teeter-totter, or an hourglass

- Photographs, maps, diagrams, and models of familiar locations, such as the classroom, school, playground, neighborhood, and children's homes

- Books that feature shapes, stories about rearranging people and objects in space, and the same scenes shown from different perspectives

Materials for measurement (KDIs 36 and 37)

- Materials that vary along one or more measurable dimensions, such as rods of different lengths, blocks of different sizes, instruments that make sounds of different pitch or loudness, beanbags of different weights, and sand timers of different durations

- Ordered sets, such as nesting blocks, measuring spoons, pillows, paintbrushes, and drums

- Ordered labels that help children find materials and return them to their storage place, such as tracings of four sizes of measuring cups or spoons on the pegboard in the house area

- Storage containers in graduated sizes

- Materials that signal stopping and starting, such as timers, musical instruments, and recording devices

- Materials that can be set to move at different rates of speed, such as metronomes and wind-up toys

- Things in the natural environment that move or change at different rates, such as slow and fast germinating seeds in the garden and insects that creep or scurry

- Unconventional (nonstandard) measuring tools, such as string, yarn, ribbon, blocks, cubes, timers, ice cube trays, and containers of all shapes and sizes

- Conventional (standard) measuring tools, such as rulers, meter sticks, tape measures, scales, clocks, grid paper, thermometers, measuring cups and spoons, and graduated cylinders

- Books that feature objects, people, and events of different size, weight, volume, loudness, temperature, speed, and duration; books in which a situation is resolved using measurement

Materials for algebra (KDI 38)

- Materials with visual patterns, such as toys in bright colors and black and white, dress-up clothes, curtains, and upholstery

- Materials to copy, extend, and create series and patterns, such as beads, sticks, small blocks in different sizes and colors, pegs and pegboards, and drawing and collage materials

- Patterned items from nature, such as shells, live plants and dried leaves, and pictures of animals with patterned fur and wings

- Original artwork and reproductions featuring patterns (e.g., weaving and basketry)

- Pattern blocks

- Signs (diagrams, photos, lists) of routines that follow patterns

- Stories, poems, and chants with repeated words and rhythms

- Songs with repetitions in melody, rhythm, and words

- Computer programs that allow children to identify and create series and patterns

- Materials that feature regular cycles and change, such as seeds that germinate at different rates and photos of children at different ages and outside at different seasons

- Books that feature poetry and recurring words, sentences, or refrains

- Catalogs with building, home decorating, and craft supplies, such as bricks, wallpaper and carpet samples, ceramic tiles, weaving and quilting patterns, and sewing fabrics

Materials for data analysis (KDI 39)

- Materials that appeal to all five senses, including items made of natural materials (e.g., wood, stone, clay), sound-making items (e.g., instruments, clocks, timers), materials with different textures (e.g., smooth, coarse; wet, dry; warm, cool; hard, soft), things that smell (e.g., flowers, herbs, spices), and foods of different tastes and textures at meals and snacktime

- Materials for collecting, sorting and labeling items, such as boxes, paper and cloth bags, plastic containers, masking tape, sticky notes, and string and pipe cleaners (for tying and bundling)

- Portable tools for recording data, such as clipboards, pencils, crayons, markers, and chalk

- Materials for diagramming or graphing data, such as large chart paper and easels, graph paper with large grids, and poster board

- Small objects to represent counted quantities such as buttons, acorns, pebbles, and plastic disks

- Books that feature collecting information to solve a problem; books about making and verifying predictions

— Epstein (2009, pp. 14, 16–17, 19, 21, & 24)

Recycled lids from plastic milk jugs are inexpensive, open-ended materials that children can use to create patterns (KDI 38).

or notice that all the pipe cleaners in the box "make the number one." Children also use all kinds of materials to measure length, such as blocks, shoes, and small toys lined up end to end. Observing the many ways that children spontaneously use materials will give you ideas to plan small- and large-group times that focus on mathematics (Neill, n.d.). For example, when one teacher heard children commenting on the "circles" they blew with bubble wands, she gave them materials to make wands of different shapes to see if they affected the shape of the bubbles.

Converse with children using mathematics words and terms

Children develop an understanding of mathematics by hearing mathematics language in their daily conversations with adults and beginning to use it themselves. Research shows a direct link between how much *number talk* children hear from their primary caregivers and their understanding of number words in preschool (Levine, Suriyakham, Rowe, Huttenlocher, & Gunderson, 2010). It is therefore important to use mathematics words and ideas when you talk with children, listen for their mathematical understandings, and encourage children to talk about their mathematics discoveries.

To introduce mathematics words and ideas as you interact with children, talk about properties, patterns, problems, and connections in the materials and actions they engage with. For example, refer to the number of people and objects ("There are only five people at small-group time today because Katie and Jim are not here"), comment on the relative amounts of things ("More children ate apple slices than pear slices at snacktime"), use measurement terms ("Your ramp is three-blocks wide" or "We need

to add one more cup of flour to make the play dough stiff enough"), and suggest using mathematical processes to answer children's questions and solve problems arising in play ("How could we measure them to find out for sure which is bigger?").

When children hear mathematics words in everyday dialogue, they *mathematize* their thinking; that is, they think about materials, actions, events, and interactions in mathematical terms. For example, they notice whether one group of objects has more or fewer than another, and later on, how many more or fewer are in the two groups. They sense that one activity lasts longer than another (work time lasts longer than planning time). Preschoolers have an intuitive grasp of these things, but when adults attach specific words to them, it helps children connect their informal perceptions to the formal concepts and processes of mathematics (Copley, 2010), as illustrated by preschool teachers in the following anecdotes:

At outside time, Kenneth tells Beth (his teacher), "My arm's too short for this jacket." "How can you tell?" she asks. He pulls up his hand and hides it in his sleeve. "The sleeve is longer than your arm," she observes. "I think the laundry did it!" Kenneth responds.

❖

"I'm older than Donald," Julia tells her teacher Sam as she arrives at preschool with her younger brother. "You're older," confirms Sam, "and you're also taller."

❖

At work time in the art area, Alana observes as she draws, "The pearl is big. Now it's very big. Now it's very bigger!" Her teacher Melody nods and says, "Yup. That's the biggest!"

In fact, research shows that the more teachers spontaneously expose preschoolers to math talk throughout the day, the greater the children's mathematical knowledge and skills (Klibanoff, Levine, Huttenlocher, Vasilyeva, & Hedges, 2006). Moreover, because the language of mathematics — symbols, numbers, shape labels, measurement units, and relational terms — is new to most preschoolers, preliminary research suggests that this is one area in which English language learners are not at a disadvantage compared to native speakers (Greenes, Ginsburg, & Balfanz, 2004).

Just as important as speaking in mathematical terms yourself is listening to what children say when they use mathematics in their play. Their comments and questions will give you insights into their knowledge and the degree to which they understand key concepts. Research shows that many early childhood teachers underestimate children's grasp of mathematical principles (Kamii, 2000). By listening to what young children reveal about their mathematical minds, you will be better able to scaffold their learning at the appropriate level, as illustrated by this interaction:

At work time in the toy area, Claire makes two piles of five bears each. She pushes the blue bears close together in a short row and spreads the red bears far apart in a long row. Then Claire tells her teacher, "There are more red bears." "How do you know there are more red bears?" the teacher asks, thinking that Claire's conclusion is based on perception (i.e., the separated red bears appear larger in quantity because

By asking this child, "What do you think will happen when you add one more cup?" this teacher helps the child use mathematical problem solving to accomplish his goal of making the liquid overflow in the bottom cup.

they are spread out). However, Claire surprises her teacher by correctly counting the number of bears in each group and then saying, "They're the same. But these [pointing to the red bears] are more long." "Oh," says the teacher, understanding how Claire is using the word more. *"There are the* same number *of blue bears and red bears, but the red bears take up* more space." *"Yes," Claire agrees with her teacher.*

Finally, encourage children to talk about their mathematical discoveries with you and with one another. Although children regularly use mathematics in their spontaneous play, they are rarely aware of this unless adults provide them with opportunities to talk about what they see and think (Zur & Gelman, 2004). Young children develop an understanding of mathematical vocabulary by hearing the words in the context of activities that interest them (discovering how many shells they collected or determining how many napkins to pass out at snacktime). Children also become more aware of using mathematical problem solving to accomplish their own goals if adults inquire about their reasoning. For example, you might ask how a child *divided* the cars so each person had the *same* amount or encourage children to predict what will happen if they *add more* red to the orange paint or *take away two blocks* from the base of their block tower.

Encouraging children to reflect on the mathematical aspects of their play helps them develop *mathematical metacognition* (Sierpinska, 1998); that is, they begin to think about and express ideas in mathematical terms. For example, if you wonder how many objects they used in their collage, children look at their pictures in terms of number as well as nonmathematical properties such as color and texture. If you comment on how tall or short a block structure is, children become consciously aware of height

as an element in their building activities. When you refer to the fact that each child is pouring a different amount of juice in his or her cup, you are encouraging children to notice volume as an attribute. Once children have this mathematical mindset, they begin to apply such words — size, nine, fewer than, longer, above — to a widening range of materials and actions as the "doctor" does in this pretend-play scenario:

At work time in the house area, "Doctor" Andrew looks at his teacher Beth's "hurt foot" and tells her, "It's 10 percent bad and 90 percent good!" "Well, what exactly does that mean?" Beth asks. "It will be all better in 10 days," the "doctor" tells her.

You can also scaffold children's thinking by encouraging them to talk to one another about their mathematical observations and conclusions. Listening to what they say will give you additional insights into how their minds are working. Moreover, when children share hypotheses and interpretations, question one another, and are challenged to justify their conclusions, they are more likely to correct their own thinking and learn from their more advanced peers (Campbell, 1999; Kirova & Bhargava, 2002). In fact, agreements and disagreements during peer dialogue more often prompt reflection and reconsideration than comments from adults (Baroody, 2000) as shown in the following two examples:

At work time in the book area, Rowan and Matthew agree to take turns with the computer mouse. Rowan flips the sand timer and takes the first turn. When she flips the timer again and continues using the mouse, Matthew protests, "Only one turn! You already turned the timer!" Rowan starts to protest but then moves aside. "Okay," she says. "Now it's your turn. But then I go again."

❖

At snacktime, Alex becomes upset when Justin, sitting next to him, breaks his cracker in half. "He got two. I only got one," says Alex. Justin pushes the two halves of his cracker back together. "No. I only have one," Justin says to Alex. Satisfied, though still looking a bit skeptical, Alex eats his cracker. When Alex takes a second cracker from the basket, he breaks it in two pieces and pushes them apart and together several times before eating those too.

Encourage children to use mathematics to answer their own questions and solve their own problems

At cleanup time, Corrin sorts markers into three piles. When her teacher asks about them, Corrin explains: "Good" [meaning works well], "bad" [meaning dry but still makes a mark], and "not at all" [meaning makes no mark]. She concludes, "The good ones have the most."

❖

At work time in the block area, Jason looks at his three paper airplanes. "This one is the biggest. This is the next biggest. Then this. Now I need to make one even littler," he says. His teacher asks how he will make it. Jason pauses and then says, "I think I have to cut the big paper in half."

❖

At outside time, Leah, Jacob, and Perez argue about who is taller. "How could you find out?" asks their teacher. "We could measure!" says Perez. Leah gets a piece of chalk and asks the teacher to mark where the tops of their heads come on the side of the tool shed. "I'm tallest," announces Jacob. "Only a little," says Perez. "When I'm five, I'll be tallest," declares Leah.

As shown in the previous three anecdotes, young children tend to solve problems intuitively and often impulsively, typically relying on habit or trial and error. However, during the preschool years, they begin to reason more logically and check out their ideas more systematically. These cognitive skills, which also draw on a child's approach to learning, can be developed through mathematics activities that encourage manipulating materials, observing outcomes, and trying to explain the results. Children will be more apt to engage in these processes — and also more willing to take intellectual risks — when the emphasis is on exploring and evaluating solutions as opposed to getting the "right answer" to the problem (Tomlinson & Hyson, 2009).

To support and encourage the use of mathematical reasoning, *do not hurry children to figure out a specific solution*. Instead, allow time for them to use trial and error and to make and verify predictions, as demonstrated by the teachers in these two anecdotes:

At work time in the toy area, Ryan and his teacher Becki are picking up paper clips with magnets. Ryan says, "I think you have more." "How can you tell?" asks Becki. "Well, I just have two and you have a whole bunch!" Ryan answers, looking unhappy. When asked how he could change who has more, Ryan says, "Take some of yours" and that's what he proceeds to do. At Becki's suggestion, they together count the number of paper clips on each magnet. "Six for me and two for you," Ryan confirms, "now I have more."

❖

Douglas and Kenneth (his teacher) are making a "seesaw" out of unit blocks. It keeps tipping to one side so Douglas adds more blocks until the seesaw lifts the other way. "There's more weight

When given the time and space, children are more likely to use mathematical reasoning to solve their problems. During cleanup time, these two children decide that the best way to carry the hollow block is over their heads.

on that side," Douglas tells Kenneth, pointing to the lower end. "That's why it's balancing now."

Observing children and letting them make their own predictions (and conclusions) allows them to self-correct their own thinking to match what they observe. Accept that children will sometimes reach a "wrong" conclusion, for example, that a seesaw tipped to one side is balanced. Their reasoning will make sense to them until later experiences enable them to adjust their thinking. Also, *beware of jumping in with an answer yourself*: "Focusing on the teacher's

reaction cuts off problem solving and reflection — not only for the one child giving the answer but also for all the children who hear the teacher's response or see the yes or no on her face. For this reason, teachers do well to cultivate an interested but noncommittal facial expression that conveys, 'Hmmm, that's an interesting idea....' and leaves plenty of time and space for children's thinking to continue" (Copley, 2010, p. 33).

It might take many repetitions before children appear to "get it," but each experience registers in their brains and allows them to gradually construct a new understanding of how numbers, shapes, units, and other mathematical phenomena work.

Pose challenges that encourage mathematical thinking

The purpose of challenging children is not to trick them, force them beyond their comfort zone, or test them to see if they come up with the correct answer. Rather, appropriate challenges can engage children in the fun of doing mathematics and, most important, foster their reasoning about whether, how, and why (or why not) the results they anticipate and observe occur.

Like appropriate scaffolding in general, mathematics challenges help children stretch their thinking without causing confusion or frustration. It supports what they already know (e.g., the number five has this many) and helps them apply this knowledge in a different or changing context (discovering how many ways you can combine objects to produce five in the set):

At greeting time, Milo tells his teacher Yasmina, "The first time you crash your car, you get 120 points. The second time you crash it, you get 220 points!" "What happens if you crash your car a third time?" Yasmina asks Milo. He thinks for a second and answers, "You get a thousand points!"

Preschoolers are generally eager to meet challenges when they originate with things that interest them (how many beads fit in the jar before they spill over the top, how many children lying end to end will stretch from the table to the bookshelf). You can further build on their natural enthusiasm by *presenting challenges as games or matters of curiosity,* rather than as tests or dares. The language you use to introduce mathematical challenges can set the

right tone for children to predict, investigate, describe, and attempt to explain (reason about) their experiences. Making comments ("I wonder what would happen if...") and occasionally asking an open-ended question ("Could you try to make it fit a different way?") are more effective than demands ("Do it this way") or closed-ended questions (e.g., "Did the smaller one work better than the bigger one?"), as demonstrated by this teacher:

At small-group time, Jonah draws a spiral that starts out large and gets smaller and smaller. Several children become very interested and ask him, "How did you do that?" Jonah shows them

Comments and Questions for Posing Mathematical Challenges

- I wonder what would happen if...
- How do you know?
- Why do you think...?
- What makes you sure?
- How could you find out?
- Perhaps it's because...
- What else can you find that works like this?
- What would it look like if you moved...?
- I wonder if you thought that would happen. Why (or why not)?
- I wonder why that happened.
- ____ tried something like that at snacktime today. What happened?
- Let's try out your idea and see what happens.
- If that's so, then it should.... Let's check it out.
- Something doesn't seem right. Let's see if we can fix it.
- What would I need to do...?
- We don't have.... What else might work?

Mathematics in Action

KDI 31. Number words and symbols

KDI 34. Shapes

KDI 32. Counting

KDI 33. Part-whole relationships

KDI 35. Spatial awareness

KDI 36. Measuring

KDI 38. Patterns

KDI 37. Unit

KDI 39. Data analysis

how to "start big and then go small and smaller." Their teacher says, "I wonder if you could start small and then go big and bigger." The children take up the challenge.

Key Developmental Indicators

HighScope has nine **key developmental indicators (KDIs)** in Mathematics: 31. Number words and symbols, 32. Counting, 33. Part-whole relationships, 34. Shapes, 35. Spatial awareness, 36. Measuring, 37. Unit, 38. Patterns, and 39. Data analysis.

Chapters 3–11 discuss the knowledge and skills young children acquire in each of these KDIs and the specific teaching strategies adults can use to support their development. At the end of each chapter is a scaffolding chart with examples of what children might say and do at early, middle, and later stages of development, and how adults can scaffold their learning through appropriate support and gentle extensions. These charts offer additional ideas on how you might carry out the strategies in the following chapters during play and other interactions with children.

After talking with his teacher about how his big sister just turned nine, this child goes over to the number book collection and selects The Number 9.

Key Developmental Indicators in Mathematics

E. Mathematics

31. Number words and symbols: Children recognize and use number words and symbols.
Description: Children recognize and name numerals in their environment. They understand that cardinal numbers (e.g., one, two, three) refer to quantity and that ordinal numbers (e.g., first, second, last) refer to the order of things. They write numerals.

32. Counting: Children count things.
Description: Children count with one-to-one correspondence (e.g., touch an object and say a number). They understand that the last number counted tells *how many*. Children compare and order quantities (e.g., more, fewer/less, same). They understand the concepts of *adding to* and *taking away*.

33. Part-whole relationships: Children combine and separate quantities of objects.
Description: Children compose and decompose quantities. They use parts to make up the whole set (e.g., combine two blocks and three blocks to make a set of five blocks). They also divide the whole set into parts (e.g., separate five blocks into one block and four blocks).

34. Shapes: Children identify, name, and describe shapes.
Description: Children recognize, compare, and sort two- and three-dimensional shapes (e.g., triangle, rectangle, circle; cone, cube, sphere). They understand what makes a shape a shape (e.g., all triangles have three sides and three points). Children transform (change) shapes by putting things together and taking them apart.

35. Spatial awareness: Children recognize spatial relationships among people and objects.
Description: Children use position, direction, and distance words to describe actions and the location of objects in their environment. They solve simple spatial problems in play (e.g., building with blocks, doing puzzles, wrapping objects).

36. Measuring: Children measure to describe, compare, and order things.
Description: Children use measurement terms to describe attributes (i.e., length, volume, weight, temperature, and time). They compare quantities (e.g., same, different; bigger, smaller; more, less; heavier, lighter) and order them (e.g., shortest, medium, longest). They estimate relative quantities (e.g., whether something has more or less).

37. Unit: Children understand and use the concept of unit.
Description: Children understand that a unit is a standard (unvarying) quantity. They measure using unconventional (e.g., block) and conventional (e.g., ruler) measuring tools. They use correct measuring procedures (e.g., begin at the baseline and measure without gaps or overlaps).

38. Patterns: Children identify, describe, copy, complete, and create patterns.
Description: Children lay the foundation for algebra by working with simple alternating patterns (e.g., ABABAB) and progressing to more complex patterns (e.g., AABAABAAB, ABCABCABC). They recognize repeating sequences (e.g., the daily routine, movement patterns) and begin to identify and describe increasing and decreasing patterns (e.g., height grows as age increases).

39. Data analysis: Children use information about quantity to draw conclusions, make decisions, and solve problems.
Description: Children collect, organize, and compare information based on measurable attributes. They represent data in simple ways (e.g., tally marks, stacks of blocks, pictures, lists, charts, graphs). They interpret and apply information in their work and play (e.g., how many cups are needed if two children are absent).

KDI 31. Number Words and Symbols

E. Mathematics

31. Number words and symbols: Children recognize and use number words and symbols.

Description: Children recognize and name numerals in their environment. They understand that cardinal numbers (e.g., one, two, three) refer to quantity and that ordinal numbers (e.g., first, second, last) refer to the order of things. They write numerals.

At work time in the house area, Jayla and four other children pretend it is her birthday. She explains to Emily, her teacher, "Today is my birthday. I am 13. My sister is 15; she is older. My brother is 16; he is older. Mom is 16 and Dad is 19."

❖

At outside time, Audie molds a pile of snow with his hands. "It takes 45 pounds of ice to build a ski mountain this big," he announces.

❖

At work time at the computer table, Jenna points to and names the numbers on the screen: 3, 4, 5, and 9. When she sees the number 11, she says, "Two number ones!"

❖

At outside time, Max uses a stick to write letters and numbers in the dirt. Some of them are written conventionally (correctly), while others look similar to the combinations of lines and curves in real symbols (letter-like and number-like forms).

How Knowledge About Number Words and Symbols Develops

Young children encounter numbers and number concepts all the time in their environment. They hear, see, handle, and learn to use cardinal number words (e.g., zero, one, two) and also work with ordinal number words (e.g., first, last, second, third). They develop what is called *number sense,* the awareness that numbers represent the quantity of things in a set and can be manipulated or operated on in various ways. Just as preschoolers begin to write letters, they also write numerals.

Cardinal numbers

Recognizing that things come in quantities (e.g., one bottle, two blocks) begins early in the second year of life. When older toddlers and young preschoolers attach a number word to a perceived amount, it is an important conceptual step in understanding that numbers represent quantity (Baroody, Lai, & Mix, 2006). Even though preschoolers do not at first associate number words with specific quantities, they begin to recognize that sets of fingers represent small amounts, for example, one finger raised means "I want one cracker" or two fingers raised means "I'm two years old." To grasp the association between number words, signs, and quantity, children must generalize from a specific example to the idea of *oneness* or *twoness*. They achieve this realization by connecting concrete objects (and, later, actions, sounds, and events) with more abstract number words and symbols.

Most preschoolers learn cardinal number words (one, two, three, and so on) by rote and can count up to 20 by kindergarten (Clements, 2004b). However, because they do not at first realize that the words represent a quantitative progression, they often say them in any order. Typically young children will say the first part of the list correctly and then omit some numbers or say lots of numbers out of order and with many repetitions (one, two, three, five, eight, two, three) (Fuson, 1988). So, instead of reciting meaningless lists of numbers, preschoolers will eventually become familiar with numerical order if adults incorporate numbers in counting songs, look at number books with children, and find natural opportunities to count with children in play ("I wonder how many acorns we found; let's count them and find out"). Combining number words with objects or actions also helps children understand one-to-one correspondence and the idea that the last number counted tells how many are in the set (cardinality) (key developmental indicator [KDI] 32. Counting).

Ordinal numbers

Although a great deal of research has been done on children's acquisition of cardinal number knowledge, less is known about how they come to understand ordinal numbers, such as first and last, and first, second, and third (Baroody, 2004). Children initially develop an intuitive or nonverbal sense of order for small collections of objects (up to three or four items) between

Developing Number Sense

Number sense refers to a child's "intuition about numbers, their magnitudes, their effects in operations, and their relationships to real quantities and phenomena" (Van de Walle, 1998, p. 4).

Children who have number sense believe that mathematics itself makes sense and can be used flexibly and practically in their everyday lives. For example, the National Council of Teachers of Mathematics (2000) notes that children who have number sense recognize the relative size of numbers and the effect of operating on them (e.g., by adding and subtracting them, or grouping and separating them in different combinations). They can also use numbers to measure common objects and events and apply numerical thinking to solve problems with materials, people, and actions. For children, number sense is as basic to learning mathematics as phonemic awareness is to learning to read (Gersten & Chard, 1999).

Preschoolers are beginning to develop number sense when they construct a notion of oneness, twoness, and so on. Young children also have an emerging concept of number when they can see the relationship of one number to another. Patricia Campbell (1999, pp. 112–113) gives the following example for the concept of (sense of) the number five, which is:

- 2 and 3
- 1 more than 4
- A lot less than 30
- Only a little less than 7
- As many fingers as are on one hand
- With 5 more, 10 altogether

An effective mathematics curriculum in preschool fosters young children's developing number sense by permitting them to understand number as more than the product of rote counting. Children develop number sense when they see numbers as representing real things, as being capable of manipulation (operations), and as being applicable to real-life situations.

the ages of 12 and 18 months, slightly later than their sense of cardinal number. At some later point, the concept of number order becomes more exact. For preschoolers, this understanding likely develops in the context of concrete experiences as they line up objects in rows or take turns playing with a toy or being the leader in a group activity.

Number symbols (numerals)

Learning to read written number symbols (numerals such as 1, 2, 3, and so on) depends on how often children encounter them in the environment and the extent to which adults point them out (Clements & Sarama, 2007). Because numerals (like alphabet letters) are standardized, children cannot "invent" the numerical system themselves. They need adults to explicitly identify and name numerals when they appear in the environment. Learning to read numerals themselves is actually fairly easy for young children. Even two- and three-year-olds can recognize some, and by age four, many children can read the numerals 1 to 10 (National Research Council, 2009).

As they learn number names and their corresponding numerals, young children also begin to write number symbols. This mathematical development parallels reading and writing alphabet letters in literacy. Preschoolers typically begin with the numerals that are easiest to draw or write (1, 3, 4, and 7) and then progress to more complex ones (2, 5, 6, 8, 9). The 0 is easy for children, too, because it is similar to drawing a circle. Preschoolers can also read and write the numeral 10.

Children will eventually learn to write numbers correctly when adults model standard numeral writing, so there is no need to correct them when they make errors (e.g., reversing numerals).

Writing numerals is a more difficult task than reading them, and many children do not master this skill until kindergarten. As with letter-writing, preschoolers often begin with numeral-like marks on a page. When they do write actual numerals, perception and motor development may limit their ability. For example, the combination of loops and curves in 6 and 9, the straight and curved lines in 2 and 5, and the crossover in 8 require complex mental visualization and fine-motor coordination. Reversing certain numerals (such as 2, 3, and 5) is also common because the left-right orientation is still difficult at this age. There is no need to correct such errors. When adults model standard numeral writing, children eventually learn how to write them correctly on their own.

Teaching Strategies That Support Using Number Words and Symbols

In addition to providing materials with numerals on them and using cardinal and ordinal number words in conversation as described in chapter 2, you can promote children's understanding and use of number words and symbols by using the following teaching strategies.

Use number words to describe everyday materials and events

Use cardinal and ordinal number words regularly as you converse with children, play alongside them, and move through the daily routine. Hearing these words in context, and in relation to the things that interest them, is more meaningful to children than merely reciting a number list. They will understand, recognize, and begin to use number words and symbols themselves when they see and hear them associated with familiar materials and recurring events. For example,

with respect to cardinal numbers, tell children at greeting time that there are two new dolls in the house area; comment at recall time that three children played in block area; observe that one person is missing from small group, and therefore the child only needs to set out seven napkins; announce that there are five more minutes until cleanup time; and sing songs with numbers in the lyrics, such as "Five Little Monkeys Jumping on the Bed."

Here are other examples of how teachers use cardinal numbers in their everyday interactions with children:

At work time in the house area, Ann Marie plays with a "baby" dinosaur. She tells Joey, "She's only eight years old." "I'm three," announces Joey. Their teacher comments, "The baby dinosaur is eight years old and Joey is three years old."

❖

At work time in the art area, Jayla talks to the teacher about her favorite television show. "It's on channel 170 and 171," she explains. "So you watch it on two channels," the teacher says.

❖

To alert the children when work time is almost over, Becki (a teacher) asks them, "What comes after five more minutes?" "Four more minutes," Petey replies.

❖

On a day when several children are absent, Andrew looks around the greeting circle and says, "There are not much kids here." When his teacher draws several stick figures on the message board with their letter links and a line through them, she counts the number absent together with the children. "Just like I said," says Andrew.

For ordinal numbers, say things like "We got on the first boot, so now let's do the second boot" and comment on the first, last, and middle items when children line up objects in rows. Refer to a numbered turn-taking list by saying something like "First it will be Jonah's turn to choose a song from the song book. The second person to choose will be Mina." In addition, when reading to children, begin with "Let's turn to the first page and see what this book is about." You can also comment when you've read the *last* page before closing the book. Finally, sing songs that use words such as *first* and *last* or add them yourself. For example, you might sing about the order children put on their clothes to play outside in the snow to the tune of "Are You Sleeping?": "First your snow pants, first your snow pants, then your boots, then your boots. Then you put your coat on, then you put your coat on, last comes mittens, last comes mittens." Here is another example of how a teacher incorporates ordinal numbers while conversing with the children:

At outside time, Pamela says, "I'm the first one out the door." Her teacher Lynette says, "I wonder who'll be the last one out the door." Pamela answers, "Nils because he always needs help putting on his boots." She stations herself by the door until Nils comes outside — last!

In addition to using English number words, you can also sign numbers and use number words in children's home languages, either on their own and/or in conjunction with their English (spoken) counterparts. For children with visual impairments, combine tactile and auditory sensations with spoken numbers. Help them feel objects or hear sounds as you say the associated number word (e.g., gently place their hand in turn on three blocks as you say "One, two, three — three blocks").

Call attention to numerals (number symbols) in the environment

Point out written numerals on equipment and materials, such as toys, tools, books, timers, job lists, and so on. For example, indicate on the message board that there will be two visitors (draw two stick figures and write the numeral 2). Point out the numerals in periodicals (e.g., the date on the cover of a magazine, the price in a newspaper advertisement). Comment that small metal race cars have numbers on them and that paintbrushes may have numbers stamped on their handles to indicate size (width). Talk about the numerals on the clock, point to and say the page numbers in a book, play board games that include spinners with numerals, and use numerals on sign-up sheets:

At work time in the block area, Bryant adds a block across the top of his structure and says, "Look. I made the letter 8!" He asks his teacher to bring over the number book and together they find the number 8. Bryant and his teacher agree that his block design looks like an 8, "only with boxes instead of circles on the top and bottom," Bryant observes.

❖

When getting up from her nap, Ellie notices the numeral 8 inside her shoe. She says to her teacher, "That's funny, my shoe says eight but I'm really four!" Her teacher looks and replies, "Your shoe has the number 8 inside, but you're only four years old."

Call children's attention to numerals by playing games that encourage them to find and name them. Together, with the children, search for numerals in the classroom, outdoor play space, school building, and neighborhood:

At large-group time, Denzel carries a cardboard 3 as he walks around the room looking for his number. He spots the numeral 3 on the wall clock, a new roll of tape with a 3M label on the box, the cover of a letter and number book, and the message board. "I'm a 3 too," he tells his teacher while holding the numeral to his chest, "because I'm three years old."

❖

At outside time, Connie and Felicia chase their teacher Victoria under the tree house and stand beside the ladder so she doesn't escape. "Hey, it's a four," says Connie, pointing to where the slope of the ladder meets the support beams.

You can also use dot-and-numeral cards so children can match a given quantity with its written symbol (e.g., two dots on the top half of the card and the numeral 2 on the bottom). For more ideas, see the letter-based games described in the companion book *Language, Literacy, and Communication* under KDI 25. Alphabetic knowledge (Epstein, 2012a) and apply them to numerals (e.g., numeral hunt or I spy numerals).

Encourage children to write numerals

Use the letter-writing strategies (KDI 29. Writing) in *Language, Literacy, and Communication* (Epstein, 2012a), and apply them to writing numerals. For example, provide a wide array of writing materials — not just markers and paper, but also sand and sticks; water and brushes to paint numerals on pavement; shaving cream; and pipe cleaners, clay, beeswax, and Popsicle sticks to shape numerals. Accept children's number-like forms as well as number-writing errors such as reversals. Encourage them to write numerals as part of pretend play (the prices on a restaurant menu) and to write or dictate numerals when they plan or recall (write the numeral 2

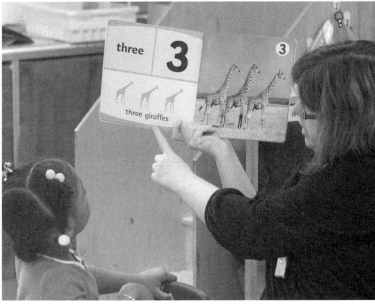

Call attention to numerals and number words in the learning environment, including those written on the message board or found in books.

under a drawing of the two towers they built at work time):

When Beth helps the children make a sign-up sheet to take turns using a new camera, several children put a numeral after their names to indicate how many turns they want.

❖

At snacktime, Crystal tells Sarah (her teacher), "Jeanette's my cousin. She's 10." She holds up 10 fingers, and says, "That's way more than me." When Crystal asks how to write the number 10, Sarah says it's a 1 and a 0. Crystal writes a 1 and asks about the 0. Sarah says it is like the letter O, and Crystal writes that next to the 1. "You wrote the number 10," says Sarah.

For examples of how children at different stages of development demonstrate their understanding of spoken and written numerals, and how adults can scaffold their learning in this KDI, see "Ideas for Scaffolding KDI 31. Number Words and Symbols" on page 37. The ideas suggested in the chart will help you support and gently extend children's understanding of number words and symbols as you play and interact with them in other ways throughout the daily routine.

At greeting time, this young girl writes the number of children (they call themselves "rainbows" because they are in the Rainbow Room) that are at school that morning: 11 arco iris (11 rainbows).

Ideas for Scaffolding KDI 31. Number Words and Symbols

Always support children at their current level and occasionally offer a gentle extension.

Earlier	Middle	Later
Children may	*Children may*	*Children may*
• Use a few number words (e.g., one, two, five).	• Use number words but not know they refer to quantity (e.g., say, "I'm three" not understanding it refers to three years or say "It costs eight").	• Understand number words refer to quantity (e.g., say, "There are three people in my family").
• Point to symbols and say number words (e.g., point to the string A37K and say, "1, 2, 3…that's my phone number").	• Recognize (read) single-digit numerals (e.g., 1, 2, 4).	• Recognize (read) several double-digit numerals (e.g., 10, 12, 25).
• Use the words *first* and/or *last* without understanding (e.g., say, "I got here first" when they arrived third or say, "We can both be first to use the bike").	• Use the words *first* and *last* correctly (e.g., say, "I'm first" before others arrive or say, "I got the fruit bowl last").	• Use a few ordinal position words (e.g., first and second) correctly (e.g., "I'll use the computer first and you can be second"; "My car came in third place").
• Write squiggles to represent numerals.	• Write numeral-like forms (e.g., 1 and 0, backward 3).	• Write two or more recognizable numerals.
To support children's current level, adults can	*To support children's current level, adults can*	*To support children's current level, adults can*
• Use number words (e.g., "You used four blocks").	• Acknowledge the use of number words (e.g., "Yes, you are three years old").	• Use number words in play (e.g., "My baby wants two more peas").
• Provide materials with numerals on them (puzzles, books, phones).	• Repeat numerals children notice (e.g., "Your race car has a 5").	• Acknowledge the double-digit numerals children say (e.g., "A 1 and a 5 are 15").
• Use ordinal number words (e.g., "The toy area is the last one left to clean up").	• Acknowledge ordinal number words (e.g., "Yes, you're the first one here").	• Use ordinal terms children know (e.g., "Dara plans second, and Paul will be third").
• Acknowledge children's interest in writing numerals.	• Ask children to read the numerals they write.	• Encourage writing numerals in play (e.g., menu prices).
To offer a gentle extension, adults can	*To offer a gentle extension, adults can*	*To offer a gentle extension, adults can*
• Use number words in questions (e.g., "Do you want one or two slices of apple?").	• Attach number words to quantity (e.g., say "You're three years old… one, two, three" while pointing to each raised finger).	• Encourage the use of number words (e.g., "How many more pegs will you put in?").
• Notice numerals during play (e.g., at the bottom of a page in a book, on a roadmap, on measuring cups and spoons).	• Point out the numerals children do not yet know (e.g., 9, 10, 12).	• Use unfamiliar two-digit numerals (e.g., "I'm turning to page 18").
• Use ordinal labels when children point (e.g., "That's the first pine cone, and this is the last one").	• Use new ordinal number words (e.g., "You're the first one. I wonder who will be second?").	• Let children continue an ordinal sequence (e.g., "Tim planned second; Ella was third. What will Pat be?").
• Provide materials for children to make numerals (e.g., play dough, sand, crayons).	• Write numerals during play (e.g., write a 2 and say "I want two pieces of pizza").	• Provide opportunities to write numerals (e.g., the number of days until the field trip on the message board).

CHAPTER 4

KDI 32. Counting

E. Mathematics
32. Counting: Children count things.

• •

Description: Children count with one-to-one correspondence (e.g., touch an object and say a number). They understand that the last number counted tells *how many*. Children compare and order quantities (e.g., more, fewer/less, same). They understand the concepts of *adding to* and *taking away*.

At the message board, Chantelle touches the two stick figures with lines through them and says, "Two kids are going to be gone sick today."

❖

At sign-up for snacktime, Joey says, "We have five kids and five jobs. Just enough."

❖

At small-group time, Jill counts the shells jumbled in her basket. She touches them at random, skipping some and touching others more than once, as she says, "1, 2, 4, 5, 3, 3, 8, 10, 100." Caitlin takes the shells out of her basket, lines them up, and then touches each shell as she counts, "1, 2, 4, 5, 6, 7, 10." Ben tells Caitlin, "Not like that. I'll show you." Turning over each shell as he counts it, Ben says, "1, 2, 3, 4, 5, 6, 7. That's how you count."

❖

At small-group time, Joshua and Ella paint small wooden blocks. Joshua counts his blocks from 1 to 17. Then Ella counts hers from 1 to 11. "I have more," Josh says.

Young children enjoy counting and comparing the number of objects, people, actions, and events in their lives. They begin to construct the rules of counting; that is, naming one and only one number for each thing counted (called *one-to-one correspondence*) and gradually learn to follow a fixed order (one, two, three, and so on). They realize that the last number they say when counting means *how many* (the total). Using manipulative objects and their fingers, preschoolers begin to understand and carry out simple mathematical operations such as *adding to* and *taking away*.

Basic Principles of Counting

More than 30 years ago, groundbreaking research by Rochel Gelman and Charles Gallistel (1978/1986) identified five principles of counting:

1. Numbers must be said in the correct sequence *(stable order).*

2. Counting involves assigning one and only one number to each thing counted *(one-to-one correspondence).*

3. The person counting must recognize that the last number said tells how many are in the set *(cardinality)*.

4. At a somewhat more advanced level is the understanding that any combination of separate items can be counted *(abstraction)*.

5. Finally, things may be counted in any order and still yield the same result *(order irrelevance)*.

The progress children make as they learn these counting principles is evident in the mistakes they make. For example, errors in counting order are common. Children typically learn the sequence from one to five but mix up the numbers that follow. Preschoolers make two types of errors in the one-to-one principle: They may point to an object and not say a number word, or, more often, they point to an object more than once or skip an object while counting (Fuson, 1988; Miller, Smith, Zhu, & Zhang, 1995).

It also takes a while for preschoolers to grasp that the last number counted represents the total. So, for example, they may count correctly but then announce a different result, as in "One, two, three, four. I have six raisins." Because they like to categorize, preschoolers may also resist counting sets composed of different things. For example, they may insist on counting beads and blocks separately, not understanding they can also count them together as one set of items.

Finally, as young children learn to count, they may think that to recount the same set, they must begin with the same item and count objects in the same order. Older preschoolers, however, are able to add on; that is, to resume counting from where they left off.

The following are typical examples of how young children count things in a preschool classroom:

At work time in the block area, Isaac correctly counts the 12 blocks he is using and then tells Sean, "I have 14 blocks." Sean counts the blocks too, skipping or double-counting some. "1, 2, 3, 5, 10, 8, 14. Yup, 14 blocks," he agrees.

❖

At work time at the computer, Trey draws a mask on the screen and counts the eyes, nose, and mouth. "One, two, three, four. Four pieces of scary guy!" he exclaims.

❖

At work time in the art area, David puts 14 staples in a piece of construction paper. After he adds each one, he counts on from where he left off after the previous staple.

How Counting Develops

Recent research shows a child's understanding of counting develops sooner than psychologists thought. For example, older toddlers can eyeball and recognize up to three items (called *subitizing*) long before they can count. Three-year-olds have a basic understanding of higher and lower numbers (Zur & Gelman, 2004). As these basic concepts take hold, children develop a more complex understanding of what counting means and how it works. Once they master the fixed order of numbers, they can say whether one number is the same as, or bigger or smaller than, another number (although very young children may consider anything more than two as "big!"):

At snacktime, Perry says, "I got two crackers and Ben got two crackers." "We got the same!" says Ben.

❖

Snacktime is one place where children's understanding of counting develops, particularly when children can "eyeball" how many cheese crackers they have and note who has more (or fewer).

At outside time, Lori points to the smaller of two piles of acorns and says, "Baby squirrels don't eat as many."

Preschoolers who can count realize that the number that comes later in the counting sequence is bigger (the *larger number principle*). They also recognize the *same number name principle;* that is, two sets are equal if they share the same number name, despite any differences in their physical appearance, as illustrated in the following anecdote:

At work time in the toy area, Hannah lines up six circles, six squares, and six rectangles on the magnet board. "They each have the same number," she says. "See, it matches."

Children also realize that counting provides information about whether two sets have the

same, more, or fewer objects in relation to one another. At first, they may simply eyeball two sets without counting, especially for groups of fewer than five objects (Sophian, Wood, & Vong, 1995), but once they have practice with counting, they apply their skills to making comparisons:

During small-group time, Jonah counts six sticks on his collage and seven pieces of foam. "There are more of those," he says, pointing to the foam pieces.

❖

At work time, in the block area, Lizzie counts her dog figures and then counts Matthew's cat figures. She says, "We both have four. We're the same."

This process (which Jonah and Lizzie used) first requires that each set be counted accurately and then that the results be compared according to which number is larger or smaller. Even more advanced is being able to say by how many one set is larger or smaller than another ("One, two, three here and one, two there. So one more in this pile"). This comparison may initially be done by lining up the two sets in one-to-one correspondence and counting the extra ones in the larger set:

At work time in the house area, Sue Ellen makes two bowls of "fruit salad" with colored beads and counts the number in each. "One, two, three, four," she counts the first bowl, and "one, two, three, four, five," she counts the second. Pointing to the second bowl, she says, "This has one more. It's for the papa bear."

❖

At small-group time, Stavros paints three circles with a line down from each of them. "Three balloons and three strings for the three bears," he tells Paulina, his teacher. Then he paints a fourth circle near the top of his page but without a line coming down. When Paulina asks him about the balloon that does not have a line, Stavros says, "That's the balloon that got away."

Later, preschoolers can use their fingers or other manipulatives to do simple addition and subtraction problems; for very small numbers, they can do the arithmetic mentally.

Even without formal instruction, preschoolers begin to develop the ability to solve such simple addition and subtraction problems. As noted previously, at first they act them out with objects (taking two blocks and adding one more). Later, when they can form mental representations (imagining two items and picturing the result with another added), they can solve simple problems in their heads. "They understand the most basic concept of addition — it is

a transformation that makes a collection larger. Similarly, they understand the most basic concept of subtraction — it is a transformation that makes a collection smaller" (Baroody, 2000, p. 63).

Preschool children also begin to construct an understanding of the base 10 pattern numbers. First they learn the names of the 10 multiples or decades, although they may confuse their order (thinking 40 comes before 30). Then they attach the one-through-nine sequence after these, sometimes miscounting beyond plus nine ("…twenty-nine, twenty-ten, twenty-eleven"):

At small-group time, Yael counts the beads on her string conventionally up to 25 and then continues counting 27, 29, 50, 60, 66. "I have 66 beads!" she says proudly. "Sixty-six beads. You put a lot on your string," agrees her teacher.

Learning this pattern is not automatic. Children depend on adults to call attention to it by using their voice to emphasize **thirty**-one, **thirty**-two, and so forth as the transition to each decade is made and by pointing to repeated numerals in a written string of numerals.

The other advance that four-year-olds make is the ability to count out a specified number of things (e.g., to count out exactly three pegs from a larger pile, to advance six boxes on a board game), as demonstrated by Chris:

At snacktime, Chris says, "There are six children and one teacher, so seven napkins." He counts out seven napkins from the basket and then puts one at each place.

Unlike with cardinality, children have to stop counting before they reach the last item to get the specified amount. Stopping can be difficult for younger preschoolers! However, being able to count "n" things is also essential for their initial exploration of addition and subtraction.

Factors That Support the Development of Counting

Developing accuracy and understanding in counting depends on several environmental factors (National Research Council, 2009). Experience is critical. Quite simply, the more counting experience young children have, the fewer errors they make. The size of the set also matters. Not surprisingly, the smaller the set, the fewer the number of errors. Even older toddlers can subitize, or eyeball, sets of up to three items before they can actually count them. Preschoolers are quite accurate counting 3 objects, then 4 or 5, eventually up to 10 or more. The arrangement of objects is relevant too. Although the grouping of three or fewer items does not matter, for larger sets, it is helpful if the objects are lined up in a row. This makes it easier for children to move their finger along, or shift the object, as they count each item. Doing an action (moving a finger or the objects) while saying a number word is a great counting aid for the physical concreteness of preschoolers. Finally, effort makes a difference (Clements & Sarama, 2007). If children are interested and care about what they are counting, they are more likely to attend to the enumeration process and strive for accuracy, as Aimee demonstrates:

At work time in the house area, Aimee uses an ice cube tray for a boat and puts one small bear in each "seat." She counts the number of bears in the first row and in the second row. "Six on top and six on bottom," she says. Then she recounts, "Just to be sure."

Parents and teachers play a critical role in helping young children understand and become fluent at counting. Early home and school experiences affect not only counting but also other numeracy skills at school entry (Benigno & Ellis, 2004). As expected, greater frequency in counting experiences is associated with higher ability. Moreover, the instructional strategies used determine the effectiveness of these experiences. Consistent with an active learning approach, development is best facilitated when adults take advantage of naturally occurring opportunities (counting the number of apples they put in the bag at the produce counter), rather than providing formal instruction in counting and its components (making children count by rote; holding up fingers and asking the child "How many?"; or quizzing a child: "How much is one plus one?"). Moreover, adults are most effective when they are sensitive to the child's level of understanding and adjust their interactions accordingly (if the child does not count with one-to-one correspondence, the adult can help the child keep track by touching each object as it is counted). When adults scaffold learning in this way and treat mathematics interactions as a game rather than a formal lesson, children are more likely to maintain their involvement, thereby increasing the opportunities for learning.

Teaching Strategies That Support Counting

As listed under general strategies in chapter 2, the development of counting depends on providing appropriate materials (discrete items children can count) and attaching vocabulary words to the process of counting and comparing (number terms and comparative words such as *smaller* and *bigger* and *more* and *fewer*). In addition, to promote young children's understanding of the different components of counting, use the following teaching strategies.

Count and compare everything

Remember that for young children, meaningful counting means *counting things,* not reciting

Meaningful counting means counting things, including other children, marshmallows, and tree stumps!

numbers by rote. Fortunately, for their own motivation, counting things is fun for preschoolers. Psychologist Howard Gardner (1991) notes that "preschoolers see the world as an arena for counting. Children want to count everything" (p. 75). Young children also seem to love the idea of big numbers (thousand, million, gazillion) even if they don't quite know what the words mean or they make them up (Ginsburg, Greenes, & Balfanz, 2003).

Being creative, teachers can invent or take advantage of many situations to count objects, actions, and events in children's daily lives. (Note that children can count objects before they can count actions and actions before they count events, Campbell, 1999). These counting activities can range from the typical (such as blocks in a tower or steps to the top of the slide) to the unexpected or even silly (such as mosquito bites on their ankles or ways to move backward). Here are some other ideas:

Notice what children count and compare, and provide similar materials and experiences

For example, provide board games with dice (moving a piece the corresponding number of places) or count the number of times a beanbag lands in a basket. Remember that rules and winning are less important to young children than counting the number of spaces or baskets. Since children like the idea of big numbers, provide opportunities for them to work with large quantities, such as counting the number of pebbles or leaves they collected on a walk or guessing the number of food pellets in the guinea pig's dish.

Use everyday activities for number learning and practice

For example, at planning and recall time, children can say how many of a certain type of material they will (or did) use ("I'm going to make a picture with all 10 markers"; "I cooked soup with four sponges"). Encourage them to gather countable materials at cleanup time (count blocks as they are reshelved) and distribute materials at small-group time (hand out one glue stick per child). Ask table setters to count how many place settings are needed. Sing and chant fingerplays using counting. Once children are adept at counting objects, you can also count the number of actions at large-group time (how many times children pat their noses).

Use and encourage children to use comparison words related to quantity

For example, at snacktime you might ask, "Julie, do you want the *same* number of crackers as Ben?" Or comment, "I was hungry! I ate three *more* apple slices than Vera." If children are absent, make a note on the message board and say, "Let's see how many *fewer* children will be here today." At recall time, say, "Let's count the clips on the recall chart to see which area had the *fewest* children." Issue simple challenges that involve comparative quantities, such as "Emma has five cars on her train. I wonder if your train will have *more or fewer* cars."

Provide materials to explore one-to-one correspondence

Provide materials that allow children to explore one-to-one correspondence such as nuts and bolts and cups and saucers. Children will also create one-to-one correspondence with any sets of materials they are playing with (put one ball of clay in each paper cup, paint a circle in each corner of the paper). Other useful materials include dot cards (also dominoes and dice) where children can match a given number of objects to the same number of dots. (Dot-and-numeral cards further help them associate quantities, number names, and numerals, as described in key developmental indicator [KDI] 31. Number words and symbols). Participating in the classroom community (KDI. 11 Community, under Social and Emotional Development) by distributing cups at snacktime also helps children develop one-to-one correspondence. Because such social activities are so meaningful to children, they are particularly effective in the development of the counting principle (Mix, 2002).

To help develop children's awareness and attach the concept of quantity to their inclination to match, comment on the numbered sets they generate ("Jason, I see you put one bear on each block. You have five bears on five blocks"). When children show an interest in counting the number of items themselves, help them keep track of what they have (and not yet) counted. Model arranging the objects in a line with clear visual separation between them. Point to, touch, or move each object aside as you and/or the child count it. Accept that children will occasionally miscount (double-count or skip as they count the items in a collection). Don't correct them; simply model the appropriate way to count. The more children hear and see you count with one-to-one correspondence, the sooner they will pick it up on their own.

Materials such as dot cards and dominoes enable children to explore one-to-one correspondence.

Engage children in simple numerical problem solving

Many opportunities arise throughout the day to pose simple challenges and identify problems that involve numbers and counting. As noted previously, the point is not to test or frustrate children but to engage them in exploring mathematical operations that are interesting and meaningful to them using the following ideas.

Pose simple addition and subtraction problems

For example, when children are ready to set the table at snacktime, remind them to look at the sign-in sheet for their small group for absences or visitors. You might say, for example, "Remember that Thomas is out sick today" or "Mrs. King is going to join us for snack." See if they subtract or add a place setting, as appropriate, and remark on the need for fewer or more utensils compared to the usual number.

Answer children's number questions

For example, if a child asks, "How old will I be on my next birthday?" count with the child and emphasize the number of years he or she is now and the number that comes next ("One, two, three, four. You're four years old now. One, two, three, four, five. Five comes after four, so on your next birthday you'll be five").

Encourage children to use counting to answer their own questions

For example, if a child says, "My daddy wants to know how many cupcakes to bring for my birthday tomorrow," you could reply, "Well, there are 16 children and two teachers. How can we figure out how many cupcakes you'll need to bring?"

Refer children to one another for help with simple number problems

For example, you might say, "Jane, you could ask Tyler to help you figure out how many blocks you'll need to make your wall one row higher." Children often face problems already solved by their peers, so, when appropriate, refer them to their peers; for example, "See how many pieces of tape Carla used to make it stick and see if the same number of pieces will work for you."

Look for stories in which counting is used to solve a problem

For example, in the book *This Little Chick* by John Lawrence, a hen hatches 10 chicks that run away. She tries to find them all. After she finds four, you might ask, "How many are still missing?"

Encourage children to reflect on their solutions

When children arrive at erroneous answers, resist the temptation to correct them. Instead, encourage them to try out their ideas and see if they work ("Benjamin says we need two more for them to be the same. Let's try that and count"). Make comments or pose questions that encourage them to rethink their solutions ("Mmm, it looks like there's still some empty spaces. How many do you think we'll need to fill them?").

For examples of how children count at different stages of development, and how adults can scaffold their learning in this KDI, see "Ideas for Scaffolding KDI 32. Counting" on page 49. The ideas suggested in the chart will help you support and gently extend children's counting as you play and interact with them in other ways throughout the daily routine.

Ideas for Scaffolding KDI 32. Counting

Always support children at their current level and occasionally offer a gentle extension.

Earlier	Middle	Later
Children may	*Children may*	*Children may*
• Rote count or "count" objects by saying numbers in random order (e.g., "Two, eight, three").	• Count up to 10 objects; may double-count or skip numbers.	• Count objects accurately (count with one-to-one correspondence).
• Recount from the beginning when asked "how many" objects.	• Say a different number than the last one counted when saying "how many" (e.g., count six objects but say there are five).	• Say the last number of objects counted tells "how many" or the total (e.g., "One, two, three, four, five. I have five grapes").
• Use general quantity words (e.g., lots, some, many, a whole bunch, a little) rather than words that compare quantity.	• Count or eyeball two sets of objects and say which has more, fewer/less, or if they are the same.	• Count or eyeball two sets of objects and say by how many one is more or fewer/less than the other.
• Add to or take away from a set of objects without understanding that the quantity of objects changes.	• Say they have more (or fewer) when they add to (or take away from) a set of objects.	• Perform simple addition and subtraction, using objects or doing it in their head.

To support children's current level, adults can	*To support children's current level, adults can*	*To support children's current level, adults can*
• Count objects (e.g., count pegs when placing them into a basket).	• Model counting objects correctly.	• Acknowledge when children count objects accurately.
• Acknowledge when children recount (e.g., "You're counting again from the beginning").	• Note the last number says how many (e.g., "One, two, three — three kids here").	• Repeat the total number, which tells how many (e.g., "Yes, there are five grapes").
• Affirm general quantity words (e.g., "You do have a lot of acorns").	• Offer matched and unmatched sets (e.g., nuts and bolts, beads and ice cube trays).	• Count and compare how many more (or fewer) objects there are.
• Use the words *adding to* and *taking away* to accompany those actions.	• Note when children say they have more (or fewer) after they add to (or take away) objects from a set.	• Add or subtract incorrectly to see if children correct the error.

To offer a gentle extension, adults can	*To offer a gentle extension, adults can*	*To offer a gentle extension, adults can*
• Count objects slowly so children can hear the correct progression of numbers.	• Recount with children by touching or moving objects while counting.	• Provide collections of more than 10 items for use in play (e.g., poker chips, stones, buttons).
• Say children are counting to figure out how many (e.g., "You're counting again to see how many buttons there are").	• Label the last number as how many (e.g., "You counted six. There are six bears").	• Ask genuine "how many" questions (e.g., "How many doggies are in this tunnel?").
• Introduce quantity words to compare (e.g., more, fewer/less, same).	• Ask children how many more (or fewer) objects there are when they compare two sets of objects.	• Provide larger sets of objects to compare how many more or fewer a set has or if they are the same.
• Note that adding to or taking away changes the quantity (e.g., "I added another bead, so now there are more").	• Do simple addition and subtraction with objects during play (e.g., "I had five cookies, I ate two, so now I have three").	• Encourage children to explain how they figured out many more (or fewer) they have when they add (or subtract) objects.

CHAPTER 5

KDI 33. Part-Whole
Relationships

E. Mathematics

33. Part-whole relationships: Children combine and separate quantities of objects.

Description: Children compose and decompose quantities. They use parts to make up the whole set (e.g., combine two blocks and three blocks to make a set of five blocks). They also divide the whole set into parts (e.g., separate five blocks into one block and four blocks).

At snacktime, Ella counts seven people at the table and two empty chairs. She says, "We're missing two people."

❖

At large-group time, Dwayne counts four empty carpet squares and says, "Four more kids to go." Then he counts the carpet squares (a dozen) with children already standing on them and announces, "Twelve kids and four kids. That's the whole class."

❖

At outside time, three children want to ride on the two-seat bus. Gillian says that Bill and Zoe can ride together. While they are riding, Gillian asks Matt to ride with her. Bill and Zoe finish their turn, and Gillian and Matt get on the bus. "Yeah," says Bill, "Us two and you two, so all four of us can ride."

❖

At work time in the toy area, Joshua makes a brown-and-black alternating pattern with small wooden cubes. When he runs out of black cubes, he takes apart the pattern and counts: "Seven blocks. Four browns and three blacks. There are more browns than blacks."

Young children explore part-whole relationships by putting together (composing) and separating (decomposing) different quantities of things. As they create and rearrange sets of things, they discover that a whole can be divided into two or more subsets (parts) and that these subsets can be recombined to make a large set (the whole). This basic understanding of how to manipulate the amount or quantity in sets of things is an important concept in mathematical operations, permitting children to use numbers with ease and flexibility.

The Meaning of Part-Whole Relationships in Mathematics

The part-whole concept is the understanding that a whole number can be made up of parts. For example, a total set of four can be *decomposed* (broken apart or partitioned) into groups made up of two and two or one and three. Likewise, groups of two and two (or one and three) can be *composed* (combined) to make a total set of four.

Understanding part-whole relationships is a precursor to performing simple numerical operations such as addition and subtraction; dividing something into equal (or unequal) shares; and, when children get older, multiplying. Young children develop ideas about composing and decomposing numbers by bringing together two other aspects of early numerical knowledge: "seeing numbers" (visualizing little numbers inside bigger numbers, such as seeing one and one and one inside three) and counting (knowing that each time you go up a number the total gets bigger and each time you go down a number the total gets smaller) (Clements, 2004b).

How an Understanding of Part-Whole Relationships Develops

Children initially develop concepts about part-whole relationships by manipulating materials, as educator Juanita V. Copley (2010) observes in her interactions with Rodney:

> Three-year-old Rodney shows me how old he is by using two fingers on the left hand and one finger on the right. He then changes hands, showing two fingers on the right hand and one finger on the left. His words

Young children learn about part-whole relationships by composing (putting together) and decomposing (taking apart) different quantities of things.

indicate that he is beginning to understand part-part-whole relationships: "See, Miss Nita, I can make 3 like this." (p. 59)

Children sort and resort sets of objects, composing and decomposing them in different ways. With experience, they can begin to visualize (represent) part-whole relationships in their minds. Most three- and four-year-olds can describe the parts of whole numbers up to five, with an understanding of larger numbers emerging around age six (Copley, 2010). The easiest are the doubles that involve equal addends (one and one make two, two and two make four), because children can see and later imagine these pairings. Even at age three, children are concerned about dividing such things as pieces of candy or the number of minutes per turn at the computer into equal shares (Clements, 2004b):

At small-group time, Warren takes several Legos from the big pile in the middle of the table. "I only need some of them," he says. Then he turns to Olin and explains, "I'm just going to use the red and yellow ones in my garage. You can have all the blue ones." He pauses and adds, "You can have the black ones too. But there are only two of those."

Next, children explore part-whole relationships that involve two groups of unequal numbers (you can make five with one and four or with two and three):

At work time in the art area, Jeff molds five play dough airplanes and inserts red beads in the front. "These three have three lights because they fly at night, and these two only have two because they fly in the day," he explains to his teacher Rodney. "So the planes that fly in the night and the ones that fly in the day have a different number of lights," Rodney says to make sure he has understood. Jeff nods and says, "Five planes, but the three that fly in the dark need more lights."

Finally, children can recognize part-whole relationships with combinations of three or more numerical groups (you can make seven with one and six, two and five, three and three and one, and so on) (National Research Council, 2009).

Children are intrigued by the notion that smaller numbers are "hiding" inside larger numbers. For example, they can take apart five to see that it can be made from a three and a two as well as a four and a one. It can even be made from a pair of twos and a one! Initially, children see the total and the parts within it. Then they shift to seeing the parts as making up (being inside) the total. For example, preschoolers first see that a set of three blocks is made up of one block plus two blocks. Later they see that one block is inside a set of three blocks and that two blocks are also inside that set of three blocks. Once they grasp this notion, children are able to *count on* (resume counting from where they left off) without having to go back to the beginning (see chapter 4). For example, they might count out a set of three objects and then decide to add more and begin counting with the next number (four) without having to starting counting again at one:

At work time in the block area, when Devon and Quay want to ride with Aaron in his spaceship, he gives them each a square hollow block to sit on. He counts out three more blocks and says, "Five seats in my spaceship — one for me because I'm the captain, two for passengers, and two left to hold the food."

❖

At outside time, Douglas rolls a ball off the deck over and over again. Each time, he counts the number of steps the ball hits on its way to the ground: "All four steps!...Two steps!...No steps!"

Research also shows that preschool children may be capable of adding and subtracting simple fractions, at least in an informal sense:

At work time in the art area, Douglas makes a pair of handcuffs with pipe cleaners, then breaks them apart. "Look, Peter," he tells his teacher, "I have two halves!"

When researchers posed an interesting challenge in a game where they hid parts of a circle, children understood that two half circles made a whole circle but that a half circle and a quarter circle made less than a whole circle (Mix, Levine, & Huttenlocher, 1999). This does *not* mean most preschoolers can — or should — be working with fractions! However, they are beginning to develop a sense of the part-whole relationships that underlie fractions.

Teaching Strategies That Support Understanding Part-Whole Relationships

To promote children's understanding of part-whole relationships, adults can use the following teaching strategies.

Provide materials that can be grouped and regrouped

Many classroom items can be grouped in various ways and then regrouped. These include small sets of toys (such as counting bears, small plastic animals, and blocks), writing and drawing materials (markers, colored pencils, and pieces of paper), and natural items (shells, acorns, and rocks). Children may divide and recombine materials for the sheer fun of moving things around (sliding them across the table or sweeping them together and shoving them apart):

To promote children's understanding of part-whole relationships, provide them with materials that can be grouped in different ways, such as counting bears and natural materials (e.g., acorns, nuts, and rocks).

At small-group time, Florio places 10 bears on a block and says, "Ten bears escaped from the zoo and took a ride on a motorcycle." He moves the block around and some fall off. He counts those that fell and says, "Five got left behind and five got to take a ride."

Or, children may have a specific category system in mind for creating two or more smaller sets from the larger one, such as subdividing or sorting the whole on the basis of color or size. Let children determine whether and how to create and re-create groups of objects:

At work time in the toy area, Carlos wants the Legos that Joey is using. Joey asks Carlos how many he wants. Carlos says he wants 8, which is half of them. Joey says, "You can have 6 and I'll have 10" and proceeds to count and divide them into two piles in those amounts.

Once children have completed making their sets, encourage them to count and compare the number of items in each. (Describing the qualitative nature of the attributes children use as the basis for sorting falls under classification in key developmental indicator [KDI] 46. Classifying under Science and Technology.)

Provide materials that can be taken apart and put back together

Many art and construction materials (including unit blocks and Legos) lend themselves to being subdivided and combined in new ways. For example, lumps of clay can be divided into small balls and reassembled back into a whole.

Likewise, toys and other classroom materials with equal-sized parts can be built up and then broken down into their component parts:

At work time in the toy area, Julia makes a series of towers with counting pegs. When she is done, she counts and reports that she used 20 pegs. Then she adds, "And look. I have one, two, three, four towers. Twenty pegs in four towers!"

As you play alongside children, comment on how they assemble and disassemble materials ("You took apart the big tower and used the Legos to make two smaller towers"). Challenge them to combine discrete materials in different ways. For example, at small-group time, distribute baskets of buttons and say, "I wonder how many ways you can make five with your buttons." Children will generally begin with two-part combinations (e.g., one and four buttons or three and two buttons). As the children gain experience with bigger numbers, you can increase the total amount (e.g., make eight) and also introduce the idea of making the total number out of more than two parts.

For examples of how children at different stages of development demonstrate their understanding of part-whole relationships, and how adults can scaffold their learning in this KDI, see "Ideas for Scaffolding KDI 33. Part-Whole Relationships" on page 57. The ideas suggested in the chart will help you support and gently extend children's understanding of part-whole relationships as you play and interact with them in other ways throughout the daily routine.

Ideas for Scaffolding KDI 33. Part-Whole Relationships

Always support children at their current level and occasionally offer a gentle extension.

Earlier	Middle	Later
Children may	*Children may*	*Children may*
• Create a set from a group of objects (e.g., pick out five beads from the basket, push together the four crayons that are on the table). • Divide a set of objects into groups (e.g., from the pile of shells, divide them into three groups).	• Combine objects into a set and describe what they are doing (e.g., gather together a set of crayons and say, "These are my favorite colors"). • Divide a set of objects into groups and describe what they are doing (e.g., separate a set of dinosaurs into two groups and say, "These are the good guys, and these are the bad guys").	• Combine objects into a set and say how many there are (e.g., pushing together plastic horses and cows, say, "I've got six animals"). • Divide a set of objects into groups and say how many there are in each (e.g., "First I had five corks. Then I made two and two and one more!").
To support children's current level, adults can	*To support children's current level, adults can*	*To support children's current level, adults can*
• Provide collections of small objects children can combine in different ways. • Imitate how children divide sets of objects (e.g., separate shells into piles alongside the children).	• Imitate how children combine objects into sets and comment on what they've done (e.g., "I made a set of my favorite color crayons"). • Repeat children's descriptions of their groups (e.g., "You made two piles. The good guys are in this one and the bad guys are in that one").	• Acknowledge that the children have created and counted a set of objects (e.g., "You put together six farm animals"). • Acknowledge that groups can be divided into sets with different numbers (e.g., "You divided the five beads so two piles have two beads and the other pile has one bead").
To offer a gentle extension, adults can	*To offer a gentle extension, adults can*	*To offer a gentle extension, adults can*
• Comment on what children do when they combine objects into sets (e.g., "You made a collection of beads"). • Describe what children are doing when they separate objects into groups (e.g., "You're dividing your shells into groups").	• Count the number of objects in the set (e.g., "One, two, three, four. There are four crayons in my pile"). • Divide the set of objects into different groups and comment on what they've done (e.g., "I made three piles with my dinosaurs — good guys, bad guys, and medium guys").	• Create a set of objects and wonder if it has the same number as the children's set (e.g., "You have six animals. I wonder if my collection has the same or a different number"). • Starting with same number of objects as the children, divide the set in a different way, and comment that the total number is the same (e.g., "We both have five. You have two, two, and one. I have two and three").

CHAPTER **6**

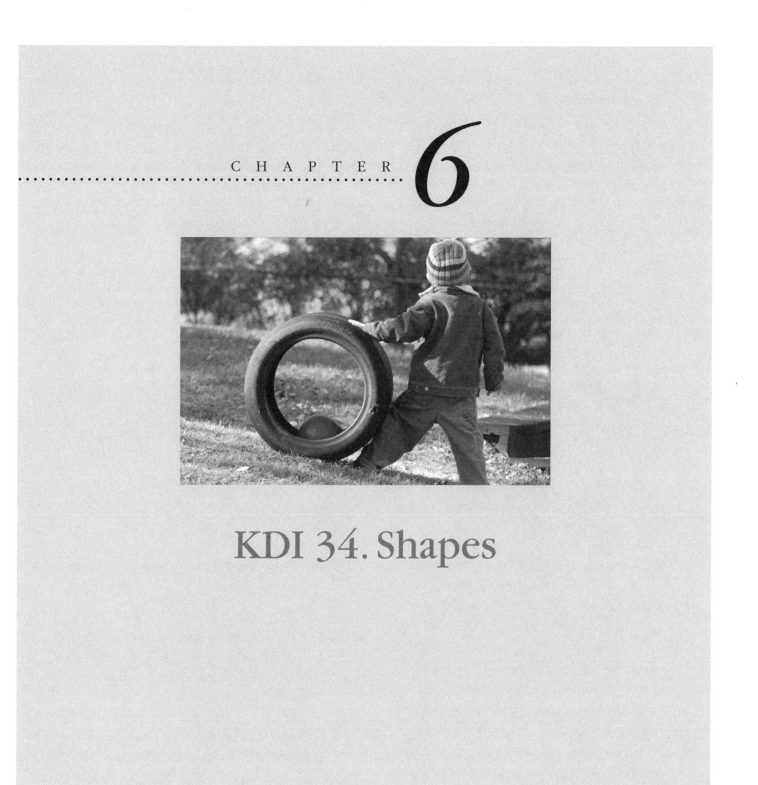

KDI 34. Shapes

E. Mathematics

34. Shapes: Children identify, name, and describe shapes.

Description: Children recognize, compare, and sort two- and three-dimensional shapes (e.g., triangle, rectangle, circle; cone, cube, sphere). They understand what makes a shape a shape (e.g., all triangles have three sides and three points). Children transform (change) shapes by putting things together and taking them apart.

At small-group time, the teacher covers a shape with a scarf and asks David to guess what it is. He runs his fingers along the sides of the hidden shape and says, "It's a rectangle."

❖

At snacktime, while eating cheese crackers, Senguele holds one up and says, "A diamond has four sides, a square has four sides. A circle doesn't have sides. It's round and round."

❖

At small-group time, Matthew uses two triangular Magna-Tiles to make "a diamond."

❖

At work time in the toy area, Joey rolls the tape on the table and says, "This is like a wheel because it's a circle."

Young children enjoy working with shapes. They handle them, for example, when they complete puzzles, and spontaneously sort shapes, for example, when they put squares in one pile and triangles in another. Preschoolers also combine and transform shapes to make other shapes. During these explorations, they often create their own shapes. Through these repeated hands-on interactions, preschoolers begin to recognize and name two-dimensional shapes (such as rectangle and circle) and three-dimensional shapes (such as cube and cylinder). They identify the characteristics of shapes (e.g., triangles have three sides and three angles). Young children also explore what happens when they combine (compose) and take apart (decompose) shapes to make new shapes.

The Importance of Shapes

Shape refers to the outline, contour, or form of objects. The young child's exploration of shape, together with emerging spatial awareness (key developmental indicator [KDI] 35), are the foundation for studying geometry. Along with number (KDIs 31–33) and measurement (KDIs 36–37), geometry is one of the three focal points identified by the National Council of Teachers of Mathematics (NCTM) as being critical to developing mathematical understanding in the preschool years. Beyond its importance in mathematics, knowing shapes serves a more general cognitive purpose. Since shape is a basic feature of the physical world, being aware of shape helps children learn the names of many things along with their attributes and functions (Jones & Smith, 2002).

Preschoolers learn about the many properties of shapes during their everyday experiences. For example, they learn that shapes may be regular or irregular; that they come in two- and three-dimensional forms; and that they have names, such as circle, triangle, rectangle, and square (an equal-sided rectangle):

Lying on her back at naptime, Abby counts the corners of the room. "Hmmm," she muses, "four corners. It must be a square." "How many corners does a circle have?" whispers Robert, lying next to her. Abby answers, "One, right in the middle." "Right in the middle," Robert agrees. Sheldon, on the other side of Abby, disagrees. "Circles don't have corners," he says.

Preschoolers discover the features that make a shape a shape, such as whether it is curved or straight sided and the number of angles it has. They transform or change shapes by combining them to make new structures and by moving them in various ways (sliding them into different positions, rotating or turning them, or flipping them over from front to back). They also explore symmetry; that is, whether or not shapes are the same on both sides when they are divided along a horizontal or vertical axis:

At work time in the house area, Bonita puts a shirt on her doll and explains to Yael, "First you do this arm and then you do this arm. See, it's the same on both sides." Yael imitates her to dress her doll. "That's right," says Bonita, "one side and then the other side."

❖

At work time in the art area, Samson folds a piece of paper in half, opens it up, and flattens it on the table. He runs his finger down the crease and says in amazement, "I made it the same on both sides!"

As with other areas of mathematical development, preschoolers begin their investigation of shape by working with concrete materials. They hold shapes in their hands and turn them in various ways. They see and identify shapes in their environment. Later, as children are able to create mental representations of shapes in their minds, they can imagine what they look

"How can I fit inside this tire?" This young child learns about the properties of circles through her everyday experiences on the playground.

like and what will happen if they undergo a transformation. For example, a younger child will physically turn the pieces of a jigsaw puzzle in the opening until it slips into place. An older child will look at the opening, pick a shape with the corresponding outline, turn it in the air, and set it in place.

How Knowledge About Shapes Develops

An awareness of shapes emerges early in cognitive development (Clements, 2004a). Children seem to have an innate ability to recognize and match shapes and do so instinctively in their play, even before they know shape names. The ability to differentiate between two- and three-dimensional shapes also appears to be intuitive. For example, young children match squares with squares and cubes with cubes rather than with one another. However, very young children are not at first able to consistently differentiate one shape from another, such as triangles from rectangles or rectangles from other four-sided shapes (National Research Council, 2009). Somewhat later, children do form general categories for shapes, but they relate them to their own experiences rather than to identifiable or distinctive properties. For example, they say a figure is a rectangle "because it looks like a door" or something is a circle "because it's like a clock" (Sarama & Clements, 2009).

The next step in development is when young children begin to analyze and describe the shapes themselves. Initially they learn about the parts of a shape (sides, edges; angles, points) and how they connect to form the closed "whole" of a particular shape (a triangle has three sides and three angles). Their emerging knowledge about number helps them focus on these defining attributes:

At outside time, James runs his hands up the side of the swing frame. He says, "Wow, this is a giant triangle!" Dionne, his teacher asks, "How can you tell?" James replies, "See, it's got two long parts that go up to a point, and the ground is the bottom."

Finally, children create an overall understanding of a shape and its parts and how the unique properties of each shape differentiate it from all other shapes. By the end of preschool, the vast majority of children can accurately name two-dimensional shapes such as circle and square, and to a slightly lesser extent, triangle and rectangle (Clements, Swaminathan, Hannibal, & Sarama, 1999):

At small-group time, Joshua explains his Magna-Tile construction to the teacher. "It's a house," he says, "a triangle and a rectangle. It [he points to the triangle] has three sides and it [he points to the rectangle] has four sides."

Children can accurately name a shape even when it varies in size or orientation; for example, they recognize a triangle no matter which one of its sides it rests on. Many also know the names of common three-dimensional shapes (ball or sphere, box or cube, cylinder) and several additional two-dimensional shapes, such as diamond and even parallelogram.

The ability to accurately name, describe, and compare shapes is an important achievement during the preschool years. It reflects the child's growing observational skills and emerging interest in classifying things based on physical attributes. Equally significant is learning how to transform shapes to bring about a desired result ("I'm making this bridge longer by adding a square block at each end and one in the middle to hold it up"). Imagining, carrying out, and describing transformations are meaningful, indicators of children's representational abilities and critical-thinking skills. Language is a vital component in all these descriptive and problem-solving activities. Thus, as the child's ability to describe shapes and changes in shape expands, so does his or her geometric understanding.

Teaching Strategies That Support Naming and Using Shapes

As described under "General Teaching Strategies" in chapter 2, children need to work directly and concretely with shapes. They also need to hear shape names and other vocabulary words associated with the properties of shapes and their transformation. To implement these and other shape-specific teaching strategies, adults can engage children in the following ways.

Provide shapes for children to see and touch

Looking at and physically manipulating shapes helps young children learn about their essential attributes. In the same way a print-rich environment exposes children to the properties of letters, so too does a shape-rich environment acquaint them with the attributes and relationships of shapes. To help preschoolers identify, describe, and compare shapes, use the following ideas.

Add two- and three-dimensional shapes to the classroom

Provide opportunities for children to explore two- and three dimensional shapes in a variety of sturdy materials, such as shape puzzles, shape boxes, wooden blocks, heavy cardboard, Styrofoam, fabric, and wooden spools:

Encourage children to feel, touch, and explore three-dimensional shapes and include more unusual shapes as well, such as cylinders and arches.

At work time in the block area, Bing walks around a circle he has made on the floor using wooden spools. "Hey," he says, "we could listen to 'Wheels on the Bus' because they go round and round."

Encourage children to run their fingers around the shapes to become familiar with their contours (curves, the number and relative length of the sides, the number and width of the angles). Make sure children encounter many examples of each type of shape so they learn to generalize its properties (e.g., include triangles in different sizes and materials and with equal and unequal angles):

At small-group time, while holding a wooden circle, Cara says, "A circle doesn't have any sides; it only has circles." Then she picks up a triangle and says, "It has three sides."

Be sure to include arches and circles as well as triangles, rectangles, and squares in sets of building materials. Also include irregular shapes, and vary the texture so children are drawn to feel as well as see the shapes.

Encourage children to draw shapes and provide models

Young children enjoy drawing shapes for their own sake or to support their pretend play (a rectangle for a "wand"). Support children's interest in drawing shapes by providing pictures, cardboard cutouts, molds, and other models of shapes, and comment on the shapes they draw and their similarity to the familiar ("Yes, a crown is round; you drew a circle"). Even if a child's shape is inexact, in the process of drawing it, the child pays attention to its contours and learns that a shape is a closed form. For example, a child drawing a rectangle uses straight rather than curved lines even if they are somewhat wobbly and the lines cross at the corners.

Encourage children to sort shapes and provide reasons for their groupings

Ask children to describe why shapes are *not* alike (how triangles differ from rectangles or how rectangles differ from squares). Encourage them to explain why something is *not* a particular shape (if it only has three points, it can't be a square; if it has three sides but one side is curved, it's not a triangle). Talking about shapes in these ways helps children understand what makes a shape a shape:

At small-group time, Imogen sorts "all the same squares" (actual squares) into one basket and "all the different squares" (rectangles) into another basket.

Encourage children to explore less common shapes

Children enjoy hearing and learning names such as cylinder, trapezoid, parallelogram, and octagon. Even if preschoolers do not immediately grasp their meaning and distinctive characteristics, they become attuned to the variety and functions of shapes in the world.

Provide materials with vertical symmetry and horizontal symmetry

Doll clothes, teeter-totters, and toy airplanes are common preschool materials that have *vertical symmetry* (left-right halves are identical) and/or *horizontal symmetry* (top-bottom halves are identical). For contrast, provide similar but asymmetric materials such as a glove, slide, and toy crane. Engage children in discussing how the two sides (or top and bottom) of objects are the same (symmetrical) or different (asymmetrical). Point out symmetry in the things children build or draw. For example, children often paint one side of a picture and then paint the same thing on the other side, such as a blue line down one side and a blue line down the other.

Use printed materials to focus on shape

Cut out photographs from magazines that feature shapes and encourage children to sort them. Create a shape scrapbook for the book area.

Encourage children to create and transform shapes and observe and describe the results

Young children naturally transform materials during play, observe and comment on the result, and may attempt to describe and explain the change. These transformations typically happen during constructive play (building structures) and often involve shapes (Chalufour & Worth, 2003). Typical examples are putting together two square blocks to make a rectangle or flattening a ball of clay into a circle. To build on children's spontaneous explorations as they transform shapes, and to encourage them to reflect on the changes they observe, use the following ideas.

Provide materials children can use to create and modify shapes

Children enjoy working with Popsicle sticks, toothpicks, marshmallows or gumdrops (or small balls of clay or play dough in place of food), and pieces of yarn. They can bend pipe cleaners into bubble wands and predict what shape bubble will emerge. As they assemble, take apart, and rearrange the materials, talk with children

During this small-group time, the teacher provides the children with pretzel sticks and marshmallows. She encourages the children to explore what shapes they can make with those materials.

about what defines the shapes they make. At large-group time, see if they can form their bodies into shapes and move to transform them:

At large-group time, Samantha lies straight on the floor with her arms pressed against her sides. "I'm a rectangle," she announces. Jonah curls into a ball and says he's a circle.

❖

At snacktime, Christian breaks his square graham cracker in half and says, "Two rectangles."

Talk with children as they create new shapes

As children combine and take apart shapes to create new shapes, encourage them to say what is the same and/or different after each change. Explore how shapes may be combined or divided to make other recognizable shapes; for example, two squares make a rectangle, a diamond splits into two triangles, two right triangles of the same size make a square or a rectangle, a circle cut in half makes two moon shapes or half circles.

Look for computer programs that allow children to manipulate shapes

Computers can be helpful as children explore transforming shapes because they are more flexible and agile than young children's hands (Clements, 1999). Appropriate programs allow children to slide, rotate, and flip shapes with relative ease and then observe the results. While computers should never replace manipulatives in the classroom, they can be a useful addition when young children want to draw, alter, and combine shapes:

At work time at the computer, Ramone advises Keisha to find the long squares, and then he will find the short squares while they play together on the shape-sorting game.

Name shapes and the actions children use to transform them

Labels help children identify shapes and the actions they perform with them; in addition, having the appropriate language at their disposal actually increases preschoolers' powers of observation and explanation (Greenes, 1999). To provide children with the words for shape names and attributes, as well as the process and outcomes of transformation, you can do the following:

Identify and label shapes throughout the environment

Apply the games you use to find letters and numerals to shapes. For example, go on a shape hunt in the classroom or on a walk. Begin with a single shape (a search for triangles) and later give children two or more shapes to find. Other options include playing shape I spy (e.g., "I see something shaped like a rectangle in the house area. What do you think it is?") or using shapes at cleanup time (e.g., "Let's put away everything shaped like a circle"). While discussing shapes with children, begin with simple labels ("On our walk, let's look for all the square signs") and gradually introduce more sophisticated ones ("You built your dollhouse with cubes"; "The shape of the stop sign is called an octagon"). Remember to supply the names of three-dimensional as well as two-dimensional shapes:

At small-group time, after Emily (a teacher) explains that the shape Matthew has is a sphere, Tasha holds one up and says, "I have a sphere." When Emily labels the shape in Ella's hand a cylinder, David finds one in his basket and says, "I have a cylinder."

Label, describe, and discuss shape attributes

Children may use different shape words, so repeat theirs and add others. For example, refer to sides and edges as well as points, angles, and

As this young child makes a pattern with the circles and rectangles, he recognizes that rectangles have straight sides while circles have curved lines.

corners. Discuss the difference between straight and curved lines. The more words children have, the more aware they become of what makes a shape a shape:

At work time in the toy area, when Barney asks Dov how to make a rocket ship out of Magna-Tiles, Dov says, "You need a triangle: three sides and three points."

Describe and encourage children to describe their actions and outcomes as they transform shapes

While children enjoy transforming materials as they build or do artwork, they typically do not reflect on the changes unless adults intentionally encourage them to do so. When talking about transformations with children, emphasize the shape they start with and the one they end up with ("You put two squares together and made a rectangle"). Provide labels for children's motions with shapes, such as sliding, turning, and flipping. Some children may be intrigued to learn more sophisticated verbs as well, such as *rotating* and *reversing*.

For examples of how children at different stages of development demonstrate their growing understanding of the names, properties, and transformation of shapes, and how adults can scaffold their learning in this KDI, see "Ideas for Scaffolding KDI 34. Shapes" on page 69. The ideas suggested in the chart will help you support and gently extend children's understanding of shapes as you play and interact with them in other ways throughout the daily routine.

Open-ended materials (such as Magna-Tiles) encourage children to create and transform shapes.

Ideas for Scaffolding KDI 34. Shapes

Always support children at their current level and occasionally offer a gentle extension.

Earlier	Middle	Later
Children may	*Children may*	*Children may*
• Manipulate shapes in play.	• Recognize and name basic shapes (i.e., circle, triangle, square); use the same label for similar shapes (e.g., call a rectangle a square).	• Identify two-dimensional shapes (e.g., rectangle, diamond, oval) and some three-dimensional shapes (e.g., cone, cube, pyramid).
• Match some shapes as they play without identifying their individual attributes.	• Identify a few shape attributes (e.g., refer to a shape's side or corner).	• Know what makes a shape a shape regardless of size or orientation (e.g., "These are both triangles. They both have three sides. This one is just longer").
• Work with shapes individually (e.g., use a circle as a "sun"); physically turn puzzle pieces until they slip into place.	• Select shapes based on their properties to make something else (e.g., put a triangle on top of a square and say they made a house).	• Combine or recombine shapes to make another specific shape (e.g., use two triangles because they want or need to make a square).
To support children's current level, adults can	*To support children's current level, adults can*	*To support children's current level, adults can*
• Provide two-dimensional materials in basic shapes (i.e., circle, triangle, square).	• Acknowledge the shapes children know and use them in conversation.	• Provide two- and three-dimensional materials in a wide variety of shapes.
• Match shapes alongside children; comment that they are the same shape.	• Affirm when children identify an attribute of a shape (e.g., "Yes, your square does have four sides").	• Encourage children to describe other attributes of the shapes they name (e.g., "It does have three sides. What else makes it a triangle?").
• Comment on how children manipulate shapes (e.g., "You flipped the circle over"; "You turned the triangle piece so it fit").	• Comment on the things children make when they combine shapes (e.g., "You made a person with a rectangle and a circle").	• Comment on the new shapes children make (e.g., "You made a rectangle with those two squares").
To offer a gentle extension, adults can	*To offer a gentle extension, adults can*	*To offer a gentle extension, adults can*
• Name basic shapes as they play with children (e.g., "I'm stacking the triangle blocks").	• Introduce the names of other shapes (e.g., rectangle, cube).	• Point out the similarities between shapes and common objects in the environment (e.g., "The tabletop is a rectangle").
• Comment on the attributes of shapes (e.g., "I'm looking for a triangle. It has three sides").	• Comment on what makes a shape a shape (e.g., "A triangle does have three sides. It also has three points").	• Ask how children know that all of a certain shape *is* that shape (e.g., "What makes all of these squares?").
• Model how shapes can be manipulated or combined to make something else (e.g., "Look what happened when I put my circle on my triangle. It looks like an ice cream cone!").	• Comment when children combine shapes and make different shapes (e.g., "You did make a kite. Those two triangles also make a diamond shape").	• Ask what shapes children used to make another shape (e.g., "What shapes did you use to make that rectangle?").

KDI 35. Spatial Awareness

E. Mathematics

35. Spatial awareness: Children recognize spatial relationships among people and objects.

• •

Description: Children use position, direction, and distance words to describe actions and the location of objects in their environment. They solve simple spatial problems in play (e.g., building with blocks, doing puzzles, wrapping objects).

At work time in the art area, Megan wraps a ribbon around a piece of paper and says, "I wrapped it around and around."

❖

At work time in the block area, Sinead stacks seven nesting boxes on top of each other to create a tower. She points to the top one and says, "This is the tippy top."

❖

At snacktime, Fernando says to his teacher, "It's my turn to sit in the middle of the table."

❖

At work time in the block area, Conway shows Sue (his teacher) how to make a dinosaur cage with a crate. He says, "You have to turn it upside down."

Young children explore space through observations and actions with objects and their own bodies. As they develop a sense of spatial awareness, preschoolers learn to attach position, direction, and distance words to describe what they see and what they do. They enjoy solving the spatial problems they encounter in play (e.g., completing puzzles, building block structures).

The Components of Spatial Awareness

Spatial awareness or spatial thinking involves two components. The first, *spatial orientation,* is knowing where you are and how to get around. It involves having a sense of your environment and the position of the objects in it, especially with respect to your own body. Young children use their emerging spatial orientation to move themselves and objects through space, as Dana and Johnny demonstrate here:

At work time in the house area, Dana lifts her feet off the floor by propping herself up on the table with her arms. This makes her appear taller. She says, "Now I'm high." She lets herself down and says, "Now I'm not." When Jill (a teacher) asks her how she does that, Dana responds, "I lift myself up."

❖

At small-group time, Johnny turns the top of his pumpkin until it fits back on exactly.

The second component, *spatial visualization,* is the ability to generate and manipulate images in your mind. It allows you to imagine and mentally move objects. For example, a young child might imagine a block of a certain shape and where it is located. A somewhat older child can mentally move that block to see if it fits into a given space without having to actually move the block first. In the following anecdotes, both Jason and Deola visualize how they are going to manipulate their materials before actually proceeding with their actions:

At work time in the art area, Jason cuts a long strip of paper, cuts a slit in it, and fits another paper strip into the slit. He calls it his "t sword."

❖

At work time in the house area, Deola draws a heart, cuts around it, and then fits it back into the paper from which she cut it. "Look," she says, "I made a puzzle!"

Preschool children make great advances in both aspects of spatial awareness — the greater their abilities in this area, the better their overall development in mathematics (Clements, 2004a).

How Spatial Awareness Develops

Spatial awareness begins at birth as newborns orient their bodies to seek the nourishment and stimulation they need (National Research Council, 2009). Infants reach for food, turn their heads to sounds, and grasp at mobiles within their view. They gaze *up* at familiar faces, *out* the window at the light, turn *over* on their side as their muscles develop, or snuggle *under* a blanket for warmth. All these bodily actions and orientations help them develop a sense of how

Spatial awareness includes spatial orientation — such as knowing that you need to step down off the block to make yourself lower — and spatial visualization — such as imagining how that piece will fit in before trying to place it in the puzzle.

space is organized in the world. At the same time, they are building a store of mental images that will eventually allow them to imagine and transform objects and their positions even when they are not immediately present.

Young children's cognitive systems are initially based on their own positions and movements. They gradually incorporate more of the external environment as they become able to move their own bodies and then to manipulate objects in space. These developments are obviously linked to the child's physical growth. However, they are also important cognitive milestones as children begin to "mathematize" their spatial knowledge. For example, preschoolers discover that size, shape, alignment, and the distance between objects affect how things appear and how they function. This understanding allows them to navigate space (move over, under, and around things), represent space (make simple drawings and maps), manipulate materials with growing skill (fold a letter to fit inside an envelope), and describe how they interact with the physical world ("We got into the car and drove across the bridge to grandpa's house").

Children as young as two implicitly use distance, direction, and their knowledge of landmarks to get about; for example, they know where the couch and kitchen doorway are and how to toddle between them. The more young children are allowed to move freely, the more readily their spatial sense develops. By preschool, children explicitly use distance and direction to reason about location and move themselves and objects through space:

At work time on the rug, Audie tosses beanbags into a cardboard box, empties the box, and then tosses all the beanbags into the box again. He moves farther away from the box with each round of tosses.

Children develop "mental maps" with landmarks along the route and have a general sense of how far it is between locations. Preschoolers can even imagine moving from place to place. For example, they can point from one location to another even if they have never traveled along that path themselves (Uttal & Wellman, 1989).

The language to describe spatial relationships is also acquired in the early childhood years and seems to develop in a consistent order, even across cultures (Bowerman, 1996). The first terms children learn are *in, on,* and *under,* which refer to position, and *up* and *down,* which refer to direction:

At recall time, Quentin describes the stairs he built with blocks and says, "I went up and down and up and down until I got all tuckered out!"

Next, children learn position words that deal with proximity, such as *beside* and *between.* Later still, children learn other position words that refer to frames of reference, such as *in front of* and *behind*:

At outside time, Casey points and says, "Look. The squirrels are chasing each other behind the bushes."

❖

At small-group time, Colin points to his pile of green and red poker chips and says, "I made a cake with grapes on top and strawberries on the bottom."

❖

At small-group time, Martina uses the glue stick all over the box she made and announces, "I glued the inside and the outside."

Distance words such as *near* and *far* also appear in preschool, although these terms are applied quite subjectively, for example, *near* might mean touching while *far* is one step away. The words for left and right are not learned until the early elementary grades.

Preschoolers also begin to build increasingly detailed mental representations of their physical environment (National Research Council, 2009). As early as age three, they can make models of spatial relationships in drawings, constructions, and simple maps (Copley, 2010). For example, they build a house with a garage next to it and a path leading up to the front door or draw a picture of the classroom with the large rug in the middle and the sink beside the easel. They can also use simple maps to help them find things such as an X marking hidden treasure in a specific area of the classroom. Older preschoolers can even interpret two coordinates (e.g., they can search for the toy that is hidden "next to the bookshelf and behind the beanbag chair"):

At work time in the block area, Kovid hides a "treasure" for his teacher Emily to find. He gives her this clue: "It's behind a straight rectangle."

In addition, as preschoolers acquire the vocabulary for spatial relationships, they can follow verbal directions to help them navigate and imagine locations, positions, and distances in space. For example, children can play hide-and-seek in which the leader tells them to look *behind* or *underneath* something. They can play a game such as Simon says in which the directions are only given verbally, without an accompanying visual demonstration. Furthermore, as children become increasingly able to take another person's perspective, they can also give directions. Thus, a preschooler can tell someone else to *move to the side* or *turn around* to imitate an action:

At work time, when the teacher is reading to three children on the beanbag couch, Marcus announces, "I want to sit in the middle." The

"Let's swing sideways. Now face me!" As children acquire the vocabulary for spatial relationships, they can follow each other's directions while playing.

children debate where "the middle" is and decide there can be "two middles." They rotate turns in those positions.

The more solid a child's spatial concepts, the more adept he or she is at visualizing or imagining movements in space; that is, a child can picture what it would be like to move him- or herself or an object without actually doing it (Clements, 2004a). Having a picture in their heads allows children to solve spatial problems with greater flexibility. For example, children can picture what size and shape block they need to complete a tower, an approach that is more efficient — and also less frustrating — than using trial and error to find the one that fits. They can look at two shapes in different orientations, for example, two identical triangles pointing in different directions, and mentally turn one around to see if they are the same shape. This spatial sense clearly enhances their ability to create concepts about shapes and their transformation (key developmental indicator [KDI] 34. Shapes).

The Importance of Spatial Awareness

The spatial sense that children develop in mathematics helps them with other areas of learning; for example, knowing about position and direction allows them to understand the direction of text (top to bottom, left to right). Likewise, learning in other domains supports the emergence of spatial concepts. For example, reflecting on why and how an artist chose a particular arrangement sensitizes children to the placement of lines, forms, and colors. An understanding of bodies and objects in relation to one another helps children learn new physical skills. Taking various perspectives is not only important in solving spatial problems but also applies to resolving social problems. And being aware of the position and movement of objects in nature can enhance a young child's construction of scientific principles. For all these reasons, providing children with materials and experiences that enhance their spatial awareness is an important component of an effective early childhood program.

Teaching Strategies That Support Spatial Awareness

To develop spatial awareness, young children need to use their own bodies and manipulate and observe objects in different configurations in the physical environment. Beginning with hands-on experiences, preschoolers also need opportunities and encouragement to mentally picture spatial arrangements and transformations. To promote these multiple aspects of young children's spatial awareness, adults can use the following teaching strategies described.

Provide materials and plan activities that encourage children to create spaces

The indoor and outdoor learning environment (and the equipment and materials in them) should enable children to create and explore spaces of all different sizes and configurations. Children develop spatial awareness when they have spaces and objects they can go over and under, inside and outside, straddle, and so on, such as empty boxes and tunnels, ramps and beams, and chairs draped with blankets. Preschoolers also need large, unrestricted areas where they can move freely as they experiment with positions, travel in various directions, and investigate distance. To help young children mathematize their explorations in space, consider the following ideas.

Provide a large space indoors and small spaces outdoors

Early childhood settings typically provide small spaces indoors and large spaces outdoors. While these are essential, children also need opportunities to experience the reverse. Therefore, to the extent your facilities permit, provide open spaces indoors such as a rug for large-group time and space in the block area to build large structures. If necessary, use movable equipment (beanbag chairs, shelves on wheels) to divide areas so they can be rearranged to create larger areas. Likewise, create cozy spaces outdoors (a grassy area inside a circle of low bushes, a corner bench). In the following two examples, the children have found both — a large space indoors to build blocks and a small, cozy space to read outdoors:

At work time in the block area, Andrew builds a large hollow-block enclosure and then lays a "floor" of unit blocks that fit tightly into the space.

❖

At outside time, Geena and Sharon park their tricycles along the side of the shed, about two feet away from the wall. They sit in the space they created and read books to each other.

Provide children with two- and three-dimensional objects in different spatial contexts

Usually when we think of spatial awareness, three-dimensional images come to mind. We picture children building with blocks, arranging dollhouse furniture, or draping a blanket over a table and crawling inside the "cave." Here are two other examples:

To the extent possible in your program space, provide open spaces indoors so that children have room to build large structures, such as railways.

At work time in the art area, Lisa stands on a large hollow block so she can reach the easel more easily. Later, at recall time, Lisa brings the block to her group and shows how she laid it flat on the floor to stand on it. Then she turns it on end and uses it as a chair for the rest of recall. "You found two ways to use that big block," her teacher comments.

❖

At cleanup time, Trey folds the picture he drew and says, "Now it will fit in my cubby."

In addition, children also explore spatial relationships with two-dimensional materials. For example, when preschoolers paint a picture, they make decisions about whether to place forms above, below, or next to other images. If they make and/or label a drawing at planning or recall, children must consider not only how

In the outdoor area, provide children with climbing structures so they can observe familiar things from unfamiliar viewpoints.

to position their depiction of materials or peers relative to one another on the page but also the boundaries of the paper itself to make sure their pictures (and words) fit.

Provide materials that let children fully and partially cover an area of space

For example, working with small tiles on a larger board (or pegs in a pegboard) encourages children to think about how much of the space they want to fill, where they will position the tiles (or pegs), and how the different elements of the composition relate to one another. Making a collage with various art materials or painting a surface that is partially covered invites the same kind of spatial reasoning:

At work time in the art area, Rachel covers her paper with paint, moves the clips that hold the paper to the easel to one side, and paints over the unpainted space where they had been.

At large-group time, challenge children to move their bodies in ways that occupy a small space (the carpet square on which they are standing) and then to move their bodies in ways that travel across a large space (the opposite side of the circle). At small-group time, provide objects and containers of different sizes so children can "wrap" presents. Younger children may put a small object inside a much larger container, while older children may seek a container in which the object "just fits." Talk with children about the position of the materials, the directions and distances they move things, and how things fit within and next to each other.

Encourage children to handle, move, and view things from different perspectives

Spatial visualization develops when children directly experience, and later imagine, the position and movement of people and objects. While some spatial relationships among objects stay the same when you move around, others change, and young children certainly do move around! Yet they rarely notice how their own actions affect spatial relationships unless adults call it to their attention. Use the following strategies to help children become aware of different viewpoints.

Encourage children to observe familiar things from unfamiliar viewpoints

For example, children can look at a table or the slide from underneath, climb the tree house to look down on the garden, put their heads between their knees to look behind them, or lie on their backs and tilt their heads backwards to look at the ceiling or different areas of the room:

At large-group time, Monty lies on his back in front of the book shelf and tips his head backwards. "The letters are upside down!" he exclaims.

❖

At outside time, Jalessa twists herself around and around on the swing and then lets the chains unwind. "I'm going twisty," she says. "It makes everything spin around with me."

Help children solve spatial problems by looking at things from different angles

Offer interesting spatial challenges that encourage children to consider people, objects, and situations from more than one perspective. Acknowledge when children use their emerging spatial sense to solve a problem, as illustrated here:

At work time in the block area, Sylvan builds two walls, one in front of the other. He gets frustrated when he pushes a long cylinder through an opening in the front wall and it knocks down the back wall. Seeing Sylvan's patience giving out, his teacher stands behind the walls and says, "I wonder what is making your back wall fall down." Sylvan says, "It's making me mad." His teacher says, "Maybe if you come around to where I am, it will help you solve the problem." Sylvan goes around to the other side and instantly sees the cylinder is longer than the distance between the two walls. He rebuilds the second wall farther back and then reinserts the cylinder in the front wall. "That does it!" he says with satisfaction.

❖

Rachel, Jenny, and Celia make a rectangular "bed" out of blocks, and all three try to climb inside. With their bodies oriented crosswise, they're too tall to fit. The girls want to rebuild the bed, but Rachel says there are no more long blocks. Whitney, their teacher, walks around the bed and says, "I wonder if there's another way you can all fit inside." Jenny gets out of the bed and follows Whitney as she circles the structure. "I know," she says, "go the other way." She directs her friends to lie down the "long way" and this time everyone fits inside the bed. "You solved the problem!" says Whitney.

Take photos of children from different perspectives during the daily routine

For example, you might photograph children building a block structure at work time, going up and down the climber at outside time, or boarding a van for a field trip. Try to get views from top and bottom, different sides, and looking inward and outward. View and discuss the photos with the children. Compare how the situation or objects look from different angles and talk about the different positions from which the photos were taken, where the children were standing, what they see from various angles, and so on:

At work time in the book area, the children look at a sequence of photographs of themselves taken at different times during the daily routine. "There's the clock," says Kenneth looking at a picture of himself at greeting circle. "And that wall was behind me."

Use books to explore spatial concepts

Many picture books depict a setting from different perspectives. For example, in one picture we might see a child's bedroom while the child is in bed. On another page, we might see the same room while a child is looking in a closet or out the window. Or, a parent might be standing in the doorway looking at the child in bed. Talk with children about these pictures, and encourage them to describe the position or distance of characters and objects in relation to one another. (Do this after children are very familiar with a book so you do not interrupt the narrative flow of the story.)

Use computers as an aid in the development of spatial concepts

Wands, touch pads, and touch screens make it easy for young children, whose fine-motor skills are still developing, to move objects on a computer display. Using these devices also makes children more aware of their actions because they must consciously decide what direction and distance to move the objects they see on the screen. Computers allow children to instantly observe the results of their actions. They can try out and compare the effectiveness of other types of movements (such as changing the direction an object travels or shifting the position of one object in relation to another). As with other areas in mathematics, computer software programs do not replace toys and manipulatives. However, they can serve as supplemental tools to enhance the dexterity and speed of children's sometimes awkward hand and body movements.

Use and encourage children to use words that describe position, direction, and distance

Communication skills are important in all areas of mathematics but perhaps no more so than in geometry. Spatial vocabulary and an understanding of how people and objects operate in space go hand in hand. Knowing various descriptive terms (e.g., on top of, next to, behind, far, near, backward, forward) helps children carry out their intentions and communicate their ideas. "Thus, it is important that young children be given numerous opportunities to develop their spatial and language abilities in tandem," says researcher and curriculum developer Carole Greenes (1999, p. 42). To support children's understanding and use of spatial vocabulary, try the following ideas.

Find and create opportunities to use position, direction, and distance words

For example, at planning and recall times, encourage children to describe what they will do (or did) with materials in spatial terms ("How will you make the pieces fit?"; "I see you moved the big blocks to the house area. How did you do that?"). At work time and group times, comment on what children do using spatial concepts ("I wonder how many of you can fit underneath the blanket"). When you play I spy, give clues using direction, position, and distance words ("I spy something pink inside the doll carriage"; "I'm thinking of something round behind the book area").

Encourage children to give directions to one another

For example, children can take turns being the leader at large-group time. In addition to showing others what movement to make, encourage them to say what they want the other children to do using position, direction, and distance words:

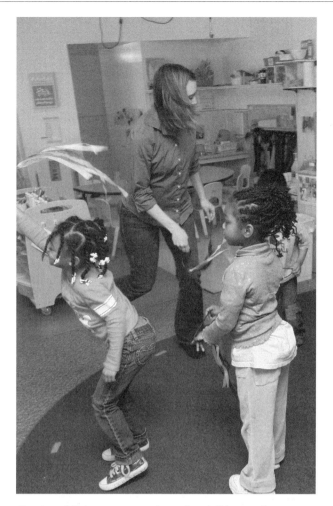

During this large-group time, the children take turns being leaders and suggest directions on how to move their ribbons. "Let's wave the ribbons way up high," says Keisha.

At large-group time, Jonetta flips a scarf over each shoulder and says she is waving them backward. *When she returns them to the front of her body, she says she is waving them* forward. *As other children imitate her movements, the teacher repeats the words* backward *and* forward *each time they shift the position of the scarf.*

As you play alongside children at work time or small-group time, ask them to explain how they did something so you can imitate them ("Tell me where on the lace I should sprinkle the sequins"). Refer children to one another to solve spatial problems ("Claire got her pieces to fit together; maybe she can tell you what she did"). Children can also give one another clues in I spy games.

Sing songs and act out favorite stories and rhymes that involve movement

Familiar songs such as "The Hokey Pokey" help children acquire and use basic spatial vocabulary such as *in, out,* and *turn around.* Other examples include "The Eensy Weensy Spider" going up and down the water spout, "The Wheels on the Bus" going round and round, "Ten in the Bed" rolling over, and so on. Encourage children to make up their own words and motions, as Desmond does here:

At large-group time, when it is Desmond's turn to lead, he says, "I know! The spider can go inside the water spout!" He buries his head inside his arms to show how he wants everyone to move as they sing and carry out his idea.

For examples of how children at different stages of development demonstrate their understanding of spatial awareness, and how adults can scaffold their learning in this KDI, see "Ideas for Scaffolding KDI 35. Spatial Awareness" on page 83. The ideas suggested in the chart will help you support and gently extend children's understanding of spatial awareness as you play and interact with them in other ways throughout the daily routine.

During this large-group time, the children sing and act out "When the Rain Comes Down." A child comes up with his own idea of what else could come down from the sky (the sun), and so the children and the teacher sing and act out "the sun shining down."

Ideas for Scaffolding KDI 35. Spatial Awareness

Always support children at their current level and occasionally offer a gentle extension.

Earlier	Middle	Later
Children may	*Children may*	*Children may*
• Move themselves or an object in response to a position, direction, or distance request (e.g., "Put the beanbag on your head"; "Can you move over to make room for Ben?").	• Describe the position, direction, or distance of an object or person (e.g., "I moved next to you so I could see the book"; "The bird is far away up in the tree").	• Compare the direction, position, or distance between people and/ or objects (e.g., "Micah is closer to the ball than Tommy is"; "The dishes are higher on the shelves than the pots").
• Continue to use the same strategy to solve a spatial problem whether or not it works (e.g., repeatedly build a tower with the largest block on top despite that it falls down each time).	• Use one or two spatial concepts to solve simple spatial problems (e.g., stand closer to the basket to throw in a beanbag after missing a few shots from farther away; prop the roof of a tent in the middle when it sags).	• Anticipate spatial problems and use spatial reasoning to solve them (e.g., write small letters on a small piece of paper; expand the hollow-block spaceship when two more children say they want to come aboard).
To support children's current level, adults can	*To support children's current level, adults can*	*To support children's current level, adults can*
• Use spatial words to help children find or place an object (e.g., "Look under the sink; maybe the sponge fell there").	• Use the spatial words that children use (e.g., "Sharifa is sitting next to me on the other side").	• Look for opportunities to compare the position, direction, and/or distances between people and/ or objects (e.g., "I'm going to bury my truck even deeper in the sand").
• Describe the spatial problem the children are having (e.g., "Each time you put the large block on top, the blocks fall down").	• Acknowledge that the children solved a problem using spatial concepts (e.g., "You stood closer so it was easier to reach the basket").	• Provide open-ended materials that create opportunities for children to solve spatial problems (e.g., blocks, collage materials, dress-up clothes, lengths of fabric, appliance boxes, funnels, and lengths of tubing).
To offer a gentle extension, adults can	*To offer a gentle extension, adults can*	*To offer a gentle extension, adults can*
• Encourage children to share their ideas using position, direction, and distance words (e.g., at large-group time, say, "Tell us where to put the beanbag next").	• Introduce new position, direction, and distance words (e.g., besides, underneath, towards, near).	• Make spatial comparisons that contrast with the ones children make (e.g., "Yes, Micah is closer to the ball and Tommy is farther away").
• Confirm when a spatial strategy is not working and ask if the children can think of another way to solve the problem (e.g., "I wonder if there's another way to stack your blocks").	• Ask children to explain why they thought their solution to a spatial problem would work (e.g., "I'm curious why you put the chair under the middle of the tent").	• Pose spatial challenges (e.g., "How do you think you could make all the furniture fit in the dollhouse?").

CHAPTER 8

KDI 36. Measuring

E. Mathematics
36. Measuring: Children measure to describe, compare, and order things.

Description: Children use measurement terms to describe attributes (i.e., length, volume, weight, temperature, and time). They compare quantities (e.g., same, different; bigger, smaller; more, less; heavier, lighter) and order them (e.g., shortest, medium, longest). They estimate relative quantities (e.g., whether something has more or less).

At work time in the toy area, Ella is stacking Unifix cubes. She says, *"Mine was only this big"* as she uses her hands to show a shorter structure. *"Now it is bigger,"* she explains.

❖

At work time in the art area, Tony comments to Eduardo about the paper planes they are making: *"My plane is heavier, so it will fly farther."*

❖

At outside time, Callie says to Andrew who is on the swing, *"I'll push you higher and faster."*

❖

At small-group time, Jon draws a large face on a large piece of paper, a medium face on a medium-sized piece of paper, and a small face on a small piece of paper.

Young children spontaneously compare quantities in general terms (same and different, more and less, longer and shorter) and estimate or anticipate quantitative differences. They grow interested in making more exact comparisons through measurement and begin to identify the attributes of things that can be measured (length, volume, weight, temperature, and time). Through repeated experiences, they begin to order objects and events according to these measurable properties.

Young Children's Interest in Measuring

For preschoolers, measurement is about realizing that things have measurable properties (such as height, volume, age, or time) and making comparisons on these dimensions (older and younger; lightest and heaviest; short, medium, and tall). They explore measuring using unconventional tools such as pieces of string and conventional tools such as rulers. As they engage in this process, they begin to understand the importance of using standard units of measurement (key developmental indicator [KDI] 37).

Preschoolers naturally encounter and discuss quantities (Seo & Ginsburg, 2004). The motivation to measure comes from their intense interest in comparing things (Copley, 2010). Young children love to compare themselves to others, for example, who is older, taller, stronger, or faster. They are also curious about comparing other people, objects, and events to one another: How many people are in my family versus your family? Is my dog bigger than yours? Is this block shorter than that one? Does the red cup hold as much juice as the blue cup? Will it take longer to drive to the park than to school? Investigating these measurement questions is one of the ways that young children make sense of their world, as illustrated here:

At snacktime, Amanda says, "My grandma's house is closer to the airport [than her house]. It's in Boston." Shane comments, "My uncle lives near the airport. It's so loud he can't hear himself think."

❖

At snacktime at the picnic table, the children suck on fruit juice Popsicles. "The longer I lick it, the smaller it gets," observes Casey. "The longer I lick it, the colder my tongue gets," giggles Kulani. The children stick their tongues out into the sun to "warm them up."

How Measurement Knowledge and Skills Develop

A young child's intuitive ideas about measurable properties — quantity, volume, weight, length, and time — are not the same as those of older children and adults (Sarama & Clements, 2009). Perceptual cues sometimes lead them astray; for example, they think a long "snake" of clay has more than the same amount of clay rolled into a ball. By preschool, some children overcome these misperceptions (they know "long and thin" compensates for "short and fat" and "conserve" the amount matter even though it changes form) while others still struggle with this logic.

Children's learning paths or trajectories in understanding measurement depend more on their experience than age or maturation, but there are some common benchmarks. For example, they recognize length as an attribute of objects beginning around age three. By age four,

children compare the lengths of two objects by matching them (holding one against the other) and can indirectly compare them using a string, stick, or some other unconventional measuring tool. They also begin to use repeated units and correct measurement procedures (i.e., using the same tool each time, starting at the baseline, and not leaving gaps or overlapping units while measuring), although they often make errors (Lehrer, 2003). With regular exposure to conventional measuring tools and units, preschoolers also develop a rudimentary understanding of length measures such as an inch, foot, or yard. (For a discussion of measuring tools and developing the concept of unit, see chapter 9.)

An understanding of area begins around age four, as children try to gauge which overall surface is "bigger" (takes up more space). Initially they do this by simply superimposing one flat surface on another. Later, they make simple side-by-side comparisons, such as covering two rectangular spaces with tiles to see which space holds more tiles. However, children at this age may still leave blank spaces, overlap edges, and go outside the boundaries. Because area involves two dimensions — length and width — it is a complex idea for preschoolers to comprehend. Not until the early elementary grades will children be able to precisely align rows and columns and count them to get an accurate measurement of area.

Children as young as two or three begin to understand volume (the amount of juice in a container, the amount of clay in a ball). They may ask, for example, "How much will it hold?" when pouring sand into a bucket or filling a tray with beads. By age four, children can directly compare the amounts in large and small containers (the amount in the smaller one fits into the larger one, the amount in the larger one spills over the smaller one). Preschoolers explore liquids and solids this way (pouring water and sand or filling a wagon with blocks). However, because volume involves multiple dimensions (length, width, and height), a full understanding of volume does not emerge until children are in the middle elementary school years.

Although understanding measurement involves making many complex connections, a growing facility with numbers helps preschoolers progress in this area of mathematics too (Clements & Stephan, 2004). For example, if children can count length or area units, or read the numerals on a scale or timer, they can compare properties more accurately and verify differences. Number sense also helps children estimate quantities more realistically; for example, they might guess that 10 crackers fit across their placemat rather than 2 or 20. However, for some attributes, such as weight and time, judgments may still be quite subjective. Something is heavier simply because it "feels heavier," and time goes fast or slow depending on whether they are bored or interested:

At work time at the computer, Jason turns and watches the sand timer while Leo sits at the keyboard. "Three times over," says Jason when the last grain of sand falls to the bottom. "It's my turn now," says Jason, and the boys switch places.

❖

At outside time, Conrad tells Nona, "You can have the bike in 20 minutes." He rides around the bike path twice, gets off, and tells Nona, "Your turn now." How many times?" she asks. "Twenty minutes," he repeats and counts as she does two laps. Then they switch places.

Repeated experiences with all these dimensions will eventually help children apply observation and reasoning to make sense of experiences that involve measuring and comparing quantity.

Teaching Strategies That Support Measuring

Opportunities to measure arise throughout the program day in an early childhood setting. For example, children may want to know how long a string they have to cut to tie a cape around their neck. If they are racing cars down a ramp, how can they tell which one is going faster? They are also typically very concerned that the cake be divided into equal pieces so everyone gets a fair share. Or they want to precisely measure how much time each child gets during his or her turn at the computer. To take advantage of these naturally occurring situations to engage children in measuring people, objects, and events, implement the following strategies.

Support children's interest in identifying and comparing measurable attributes

As noted previously, children's interest in measurement derives primarily from an interest in making comparisons. Provide materials and experiences that initially enable them to compare things on length, such as longer and shorter, wider and narrower, higher and lower, and how far around (circumference). Then provide opportunities to measure and compare area (covers more or less), volume or capacity[2] for solids and liquids (holds more or less), weight (heavier or lighter), temperature (hotter

---

[2]The word *volume* is sometimes used to refer to solids and *capacity* to refer to liquids (Copley, 2010). However, these terms are also used interchangeably in the literature.

or colder), sound (louder or softer), and two aspects of time — duration (lasts longer or shorter) and speed (goes faster or slower):

At work time at the sink, Brandon fills a sponge with water and squeezes it out several times. His teacher Sam picks up a sponge and imitates him. "It's heavier when it's wet," Brandon tells Sam as he dunks his sponge in the water. "Now make yours lighter," he says to Sam. Sam squeezes out his sponge. "There you go!" says Brandon and squeezes the water out of his sponge too.

❖

At large-group time, Andrew advises Dylan, "Put the cups inside one another to make it noisier." Dylan tries Andrew's ideas and is pleased with the sound of the smaller metal cup rattling around inside the bigger one.

Understanding the measurement of area (which involves two dimensions) and volume (three dimensions) can be difficult for preschool children. However, while they may have difficulty with the measurement process itself, they can readily compare things nonquantitatively. For example, when investigating area, children can easily see when the edges of one surface overlap another. Likewise, when exploring volume, they can observe when something is partly empty, full, or overflowing. Therefore, experiences comparing area and volume are valuable as a preliminary to later being able to quantify the differences that children observe.

After children have many experiences comparing two items, they can begin to compare three or more items and put them in order. For example, they compare length from shortest to longest, volume from holds the most to holds the least, or speed from slowest to fastest. Children may also enjoy comparing and

This young girl learns about measuring time while waiting to use the swing. Once the sand in the sand timer goes from the top bottle to the bottom one, it will be her turn on the swing.

ordering such things as sounds (higher to lower pitch [musical note], softest to loudest noise) and colors (dullest to brightest, pink to medium red to dark red):

At small-group time, Penny lines up Cuisenaire rods in order of size and says, "Little ones, medium, large, and extra large ones."

❖

At work time in the book area, Ella points to the puzzle with seriated chickens while saying, "Big, large, medium, almost tiny, almost tiny, almost tiny, tiny."

There are many materials that children can use to discover what attributes can be measured and compared. Chief among these are construction materials (blocks, boards, art supplies) and things for filling and emptying (water and sand). Children also compare their own attributes (height, age, speed, hand size), family characteristics (whose dog is bigger), the temperature or sweetness of the foods eaten at snacktime, and how much effort it takes to carry a basket of feathers versus a basket of beads (an indicator of their relative weights):

Eileen, a teacher, tells the children that tomorrow is her birthday and she will be 40 years old. "My grandpa is really old too," Zeke tells her. "Like 100 years, I think."

❖

At work time in the house area, Roseanne hands Leona a cup and says, "Here's your hot chocolate, dear. It's very, very hot. You have to wait two hours until it cools off." Leone sits patiently for about 15 seconds until Roseanne tells her it is cool enough to drink.

Everyday experiences also provide opportunities for measurement. For example, a clock can be used to measure how much time is left before cleanup begins. A field trip to the pumpkin patch is a chance to compare the capacity of different vehicles to hold adults and children as well as to compare the weight of the pumpkins children carry home. A "race" between two children from the tree to the table is an opportunity to count seconds to see which child gets there in less time.

Social conflicts also lend themselves to measurement-based solutions. For example, children may use a sand timer to make sure the duration of turns at the computer or playing with a toy is equal. Or they may measure the length of the rug so each child has the same amount of space to play.

Encourage children to estimate quantities

As children measure and compare attributes, ask them to estimate relative quantities; that is, to say which has more or fewer/less, which is heavier or lighter, and so on. Then encourage them to measure to verify their estimates. Remember that it does not matter if their predictions are right or wrong. The point is that estimating helps children reflect on the attributes they are measuring.

In addition to making estimates, children need opportunities to explain their reasoning. For example, you might say, "I wonder why you think the red bowl holds more than the blue bowl." Encourage children to compare their estimates and the thinking behind them. Often children will learn more from these discrepancies — and challenging one another — than from any comment an adult makes. After children measure to verify their estimates (e.g., counting the number of cups of water it takes to fill the two bowls), ask if the result matches their predictions. Again, encourage them to explain why (or why not) the outcome was what they thought it would be. Encourage children to listen to the explanations offered by their peers and, if necessary, help explain their reasoning to one another ("Brian thinks there will be more blue beads in the jar because they are smaller. Josie says the red ones stick closer together [they are straight sided] so more of them are crammed in. Let's count. Mmm — more blues beads than red. I wonder why.")

Use and encourage children to use measurement words

Young children benefit from hearing measurement terms and being encouraged to use this vocabulary themselves. As with other new words, children may need to hear them in

While exploring Cuisenaire rods with a child, this teacher uses measurement words to talk with the child about the different heights of the rods.

context many times before they understand and can use them correctly themselves. However, having the appropriate language helps them focus their observations and describe their actions. In this way, vocabulary can actually facilitate the development of the underlying measurement concepts.

There are three categories of measurement words that children need to encounter as you converse with them throughout the day. First are terms for *attributes* that can be measured, such as length, height, and time. Second are *comparison words* (to make two-way comparisons and also to seriate three or more items), such as longer, shorter; fastest, slowest; and tall, taller, tallest. Third are terms for common and familiar *measurement units*, such as inches, days, and

teaspoons. Here are some additional vocabulary words to introduce and use in different measurement domains:

- Length terms — length, long, tall, short, wider, narrower, width, inch, foot, ruler, yardstick (and for distance as a measure of length between two points: near, far, closer together)

- Capacity and volume terms — volume, empty, fuller, holds more, holds fewer or less, pint, quart, gallon, liter, measuring cup, measuring spoon

- Weight terms — weight, heavy, lighter, heft, scale, balance, ounce, pound, ton

- Time terms — time, speed, fast, quicker, slower, soon, later, last longer, minute, hour, day, year, clock, timer, stopwatch

During this large-group time, the teachers use (and encourage children to use) measurement words such as higher, lower, faster, *and* slower *to describe how they are moving both the parachute and the sponge balls.*

- Temperature terms — temperature, hot, heat, warmth, cool, colder, freezing, melting, thermometer, degree

  For illustrations of how children at different stages of development demonstrate their understanding of attributes and comparisons in measuring and how adults can scaffold their learning in this KDI, see "Ideas for Scaffolding KDI 36. Measuring" on page 93. The ideas suggested in the chart will help you support and gently extend children's understanding of measurement as you play and interact with them in other ways throughout the daily routine.

Ideas for Scaffolding KDI 36. Measuring

Always support children at their current level and occasionally offer a gentle extension.

Earlier	Middle	Later
Children may	*Children may*	*Children may*
• Use items without regard to the measurable attributes described (e.g., set the table with small and medium plates when another child says, "Let's set the table with small plates"). • Use a measurement term to describe a single person, object, or event as part of its identity (e.g., "I'm big"; "I'm a fast runner").	• Describe measurable attributes based on perception such as lifting or eyeballing rather than measuring (e.g., pick up a block and say, "This is heavy"; look at a cup and say, "I have lots of juice"). • Use "er" measurement words to compare two people, objects, or events (e.g., taller, shorter; hotter, colder; faster, slower).	• Use measurement to compare the attributes of people, objects, and events (e.g., align two dinosaurs and say, "Yours is longer"). • Use "est" measurement words to compare or order three or more people, objects, or events (e.g., from shortest to tallest or highest to lowest).
To support children's current level, adults can	*To support children's current level, adults can*	*To support children's current level, adults can*
• Provide materials that vary along basic measurable attributes (e.g., streamers of different lengths, sealed plastic bottles with light and heavy fillers). • Affirm the measurement words children use (e.g., "That is a big tower!"; "You did run fast").	• Talk with children about the measurable attributes they identify (e.g., "That block does feel heavy"). • Make comments and ask questions using "er" words (e.g., "I wonder which brush is wider"; "The big board was heavier so it took two of you to carry it to the house area").	• Label children's measuring actions as *measuring* (e.g., "You put the two dinosaurs together to see which was longer — you measured them"). • Provide materials in sets of three or more that lend themselves to seriation (e.g., blocks, beads, paintbrushes, doll clothes).
To offer a gentle extension, adults can	*To offer a gentle extension, adults can*	*To offer a gentle extension, adults can*
• Comment on basic measurable attributes while playing with children (e.g., "You put the long block on the bottom"; "These feathers are so light, they float up to the sky"). • Repeat children's measurement words; supply opposite words (e.g., "It is big! I'm going to look for a small one").	• Ask children how they reached their conclusion (e.g., "How do know this one is heavier?"). • Add a third item to the two children are comparing (e.g., hand them a third block of a different size and say, "You've got a smaller block and a bigger block. I wonder where this one goes").	• Model measuring to compare things on a variety of attributes (e.g., align two dinosaurs and say, "Yours is longer but mine is taller"). • Model and comment on reversing the order of seriated sets (e.g., "Suppose I put the tallest rod at this end. Where would I put the rest?").

CHAPTER 9

KDI 37. Unit

E. Mathematics

37. Unit: Children understand and use the concept of unit.

Description: Children understand that a unit is a standard (unvarying) quantity. They measure using unconventional (e.g., block) and conventional (e.g., ruler) measuring tools. They use correct measuring procedures (e.g., begin at the baseline and measure without gaps or overlaps).

Darla asks her teacher Mark, "How many tall am I?" "How could we find out?" he wonders. Darla thinks for a minute and gets a plank from the block area. She stretches out on the floor so Mark can measure her. "One board and a little bit more," he shows her. Then Mark stretches out on the floor so Darla can measure him. "One and a lot more," she announces.

❖

At work time in the house area, Matthew explains to his teacher Emily how to make a pie. "Put the pie in the oven for seven minutes," he says. He moves his finger as he counts off each section of the dial on the timer. When he reaches the seven, he says, "Then you take it out, but don't burn yourself, and then you eat it."

❖

At small-group time, Angela makes hash marks on a piece of paper as she counts how many cups of water it takes to fill the bowl "all the way to the top, but not too much." When the bowl is nearly full, she counts the marks and says, "Fifteen cups. That's a big bowl!"

Repeated experiences with measurement help children develop a concrete understanding of what a measurement unit is. Young children use both unconventional tools (such as blocks and paper cups) and conventional tools (such as rulers and measuring spoons) as their measurement unit. Gradually, they begin to understand how units function; that is, you must use the same unit to measure something and you can't switch to a different unit midway. Preschoolers also begin to grasp and use the basic principles of measuring with units — you must start at the baseline (not someplace in the middle) and you have to measure without gaps (spaces between units) or overlaps (duplicating all or part of a unit).

The Importance of Unit in Measurement

The understanding and use of *unit* is essential to accurate measurement. A unit is the standard quantity by which measurement is expressed. For example, length may be measured in inches, weight in pounds, volume in teaspoons, and length in days. The two concepts children need to learn about unit is that measuring involves repeating equal-sized units and that one counts the number of units to arrive at the total quantity (Sophian, 2004). Most young children grasp these principles by the end of preschool. They typically start measuring length by laying multiple copies of the same-sized units end to end, for example, a row of long blocks to measure the length of the room (Clements, 2004b). However, they may think they need as many tools or units as the length they are measuring:

At work time in the toy area, Dawn plays at the small Duplo table and says, "Hey, it takes more than five of the long ones to go across this table!"

It does not occur to children — until an adult suggests it — to mark the end of one unit and then lift and reposition the tool to continue measuring (Clements & Stephan, 2004).

How an Understanding of Unit Develops

Children need general measuring experiences (key developmental indicator [KDI] 36), especially making comparisons, before they can grasp the idea of unit (Copley, 2010). First they need to recognize that something is longer (or shorter) or has more (or less/fewer) than a comparable object. Next children need to connect number to these comparisons; that is, they have to recognize that one item is X amount or quantity bigger or smaller than another item (e.g., "This row is *five* shells longer than that row").

Only after children understand that comparisons between items can be made on the basis of measurable quantities are they ready to determine what those quantities are by using appropriate measuring tools:

At work time in the toy area, Marta comments, "It takes all the blue and red links to get from the steps to the toy shelf, and you have to add the green links to get to the water table."

Children can do this by using unconventional (nonstandard) measuring devices (e.g., a shoe [for length], an unmarked paper cup [for volume]). Correspondingly, children may use conventional (standard) tools (e.g., a ruler, a measuring cup). Although educators have long advocated that children progress from using unconventional to conventional tools, recent research indicates that they do equally well, and may in fact prefer, to use conventional tools from the outset (Lehrer, 2003). Materials that strike a balance, such as unit cubes, are effective in transitioning children between these two types of measuring devices.

Whatever measuring tools children use, however, it is clear that the concept of unit begins with concrete experience. Children need to manipulate both the thing being measured and the tool they are using to measure it. This helps them develop *unit sense* in the same way that working directly with numbers allows them to construct *number sense* (Clements & Stephan, 2004):

At work time in the block area, Billy says to Morgan, "It's a long time until recall." Morgan nods. Their teacher ponders, "I wonder how you'll know when it's almost recall." Billy points to the wall clock and says, "When that long thing points to a number, it's five more minutes." "Five minutes," agrees Morgan.

Often it is measurement surprises — coming up with a different total when they measure the same thing a second time or arriving at a different measurement result than someone else — that helps children discover the basic principles of using units to measure. These principles (or procedures) are:

1. Use the same unit each time.
2. Begin at the baseline (called *zero point*).
3. Measure without overlapping units or leaving gaps between units.

Children gradually discover that when they don't follow these procedures, they cannot depend on the accuracy of their answers.

Offer children both conventional measuring tools (e.g., a tape measure, a measuring cup) as well as unconventional ones (e.g., a turkey baster, a funnel).

Experience with a variety of measuring devices also helps children learn that changing the size of the unit changes the result. That is, using shorter units results in a larger total quantity (number), while using a longer unit produces a smaller total quantity. The same applies for other measures such as volume or capacity. For example, if a block is the unit, it takes fewer larger blocks than smaller blocks to fill the wagon; if a cup is the unit, it takes more smaller cups than larger cups to fill the bowl. With adult guidance, older preschoolers can also see the effects of combing smaller units to make a large unit. For example, children who regularly use unit blocks realize that two short blocks equals one long block. They may also choose a longer unit to measure something long, such as a slide, and a shorter unit to measure something short, such as a book:

At work time, while waiting for his turn at the computer, Joshua notices that the timer has run out. He says to Matthew, "Wait, you can have a longer turn. I'll use this little timer." Joshua turns over a smaller timer to give Matthew more time at the computer before it's his turn.

Teaching Strategies That Support an Understanding of Unit

Because of their propensity to differentiate and order things, children often find themselves comparing quantities. Adults can take advantage of children's inherent interest in making accurate comparisons to promote their understanding and use of unit in the measurement process. The following teaching strategies will help you address children's curiosity and scaffold their emerging knowledge about unit.

Support children's use of conventional and unconventional measuring tools

Children can measure using both nonstandard or unconventional tools ("My napkin is five pretzels long") and standard or conventional tools ("My Lego spaceship comes to the 5 on the ruler"). As noted previously, many preschoolers actually prefer the latter, at least when it comes to length (Lehrer, 2003). Unit-based toys (such as unit blocks, unit cubes, and Cuisenaire rods) are an effective bridge between the two systems. Whichever tools children use, however, it is their investigations that will help them construct an understanding of the role of units in measurement.

Although children are inherently curious about measurement and comparisons, sometimes you will need to more actively engage them in the exploration of units as a component of measuring. This can be done as a small-group activity. Unconventional tools may help in this process because they are novel and quirky. For example, it is fun and silly for preschoolers to measure how many shoes long their teacher is compared to each child in their small group or to use a turkey baster to measure water:

At work time at the water table, Alana experiments with the turkey baster, figuring out how to fill it with water for a "long squeeze." She explains, "Then I won't need so many."

Unconventional tools are also familiar and comfortable. For example, children can connect something familiar to a new process by counting how many repetitions of a favorite recording it takes for them to clean up after work time. To supplement the measurement tools you provide, encourage children to come up with their own measuring devices. They can be as creative as they like. Ask them why they think something would (or would not) be a good tool to use for measuring the attribute they are interested in.

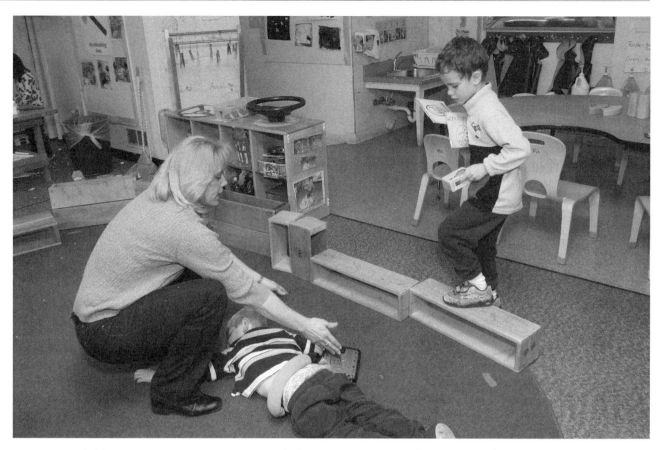

Encourage children's creativity in coming up with their own measuring devices. Here the teacher supports the children's idea of using the large hollow blocks to measure Leo.

Model accurate measuring techniques

As noted previously, there are three basic principles that children need to learn to measure: (1) Use the same measuring tool (unit) each time, (2) begin at the baseline (also called the *zero point*), and (3) measure without leaving gaps or overlapping units. Measuring accurately does not come naturally to children. They learn by seeing adults measure and explain what they are doing. Therefore, it is important to demonstrate and describe the measuring techniques you use.

When children do not use measurement principles (e.g., leaving gaps when they move the string or ruler to the next place), do not correct them. Instead, measure correctly and explain what you did. Take advantage of situations where the children themselves measure the same thing and are surprised when they come up with different results. Encourage them to resolve the difference with one another, for example: "Joshua says the table is two boards wide and Eliza came up with three boards when she measured. I wonder if you two can figure out why your totals aren't the same."

For examples of how children at different stages of development demonstrate their understanding of unit and how adults can scaffold their learning in this KDI, see "Ideas for Scaffolding KDI 37. Unit" on page 101. The ideas suggested in the chart will help you support and gently extend children's understanding of unit as you play and interact with them in other ways throughout the daily routine.

Ideas for Scaffolding KDI 37. Unit

Always support children at their current level and occasionally offer a gentle extension.

Earlier	Middle	Later
Children may	*Children may*	*Children may*
• Use measuring tools in their play (e.g., make a road with rulers, fill and empty both sides of a balance scale, use measuring cups as scoops at the water table).	• Compare things directly (e.g., hold one stick against another to see which is longer).	• Use unconventional or conventional measuring tools to answer questions that arise in their play (e.g., see how many shoe lengths it is from the door to the bookshelf, see how many measuring cups of sand a bucket will hold).
• Imitate using measuring tools without an awareness of measuring (e.g., stretch a tape measure along a row of blocks because they've seen someone else stretch and retract the tape).	• Begin to measure things but not use standard procedures (i.e., change the unit, begin measuring in the middle, overlap units or leave gaps); for example, partially fill a cup when measuring how many cups it takes to fill a bowl.	• Measure using standard measuring procedures most of the time (i.e., use the same unit, begin measuring at the baseline, do not overlap units or leave gaps).
To support children's current level, adults can	*To support children's current level, adults can*	*To support children's current level, adults can*
• Provide the names for measuring tools (e.g., "You made a road with rulers"; "You put the beads in the scale").	• Acknowledge and describe children's direct comparisons (e.g., "You found out that the red stick is longer than the blue stick by putting them next to each other").	• Acknowledge when children use units to answer their measurement questions (e.g., "You used a shoe to measure the distance from the door to the bookshelf"; "You found out the bucket holds 10 cups of sand").
• Imitate children's actions and describe what they are doing (e.g., stretch out the measuring tape and say, "We're both stretching the measuring tape"; copy children by putting beads in the balance scale).	• Model correct measuring procedures (i.e., use the same unit each time, begin at the baseline, leave no gaps or overlaps).	• Comment on children's use of correct measurement procedures (e.g., "You started measuring at the bottom of both towers so you could tell which was longer when you got to the top").
To offer a gentle extension, adults can	*To offer a gentle extension, adults can*	*To offer a gentle extension, adults can*
• Model comparing things directly (e.g., "I'm putting the tubes next to each other. I'm trying to figure out which one has more water").	• Provide a variety of measuring tools; comment on the units in the tools the children use to measure (e.g., "This is called a ruler and these marks are called inches").	• Ask children if there is another way they can measure to compare (e.g., "How else could we measure to see how much taller I am than you are?").
• Refer to children's actions as measuring (e.g., "You're weighing the beads").	• Describe inconsistencies in children's measuring procedures (e.g., "You want to see how many cars long your road is. I see you left a space between these two cars").	• Make measurement errors and see if children correct you; if not, measure again correctly while describing the correct way to measure.

CHAPTER 10

KDI 38. Patterns

E. Mathematics
38. Patterns: Children identify, describe, copy, complete, and create patterns.

Description: Children lay the foundation for algebra by working with simple alternating patterns (e.g., ABABAB) and progressing to more complex patterns (e.g., AABAABAAB, ABCABCABC). They recognize repeating sequences (e.g., the daily routine, movement patterns) and begin to identify and describe increasing and decreasing patterns (e.g., height grows as age increases).

At small-group time, Emily (the teacher) starts a pattern with beads: green, orange, green, orange, green, orange, green. When she stops, Eli hands her an orange bead from the pile to continue the pattern.

❖

At large-group time, Millie makes up this pattern: "You rub the top of your head and then you rub your stomach." She does it twice and then the other children copy her.

❖

At the message board on Friday morning, Sally Ann "reads" the picture of two crossed-out schoolhouses, "No preschool for two days." "Oh boy! Just like last week," Quentin sighs.

❖

At work time in the toy area, Tasha makes alternate lines of yellow and green tiles going up and down a small wooden ramp. "Look," she says, "I made a pattern!"

Patterns appear everywhere — in objects, movements, sounds, activities, and events. Young children become aware of these patterns through their own observations and when adults call their attention to the patterns they encounter in play. Preschoolers learn to recognize, describe, copy, extend, and create simple patterns (e.g., ABABAB) and progress to working with more complex patterns (e.g., AABAABAAB, ABCABCABC). Young children can also identify and describe patterns or changes that increase in quantity (children get bigger as they get older) and that decrease in quantity (a bowl of cereal gets smaller by one spoonful at a time).

Children's Early Experiences With Patterns

Preschool children are already aware of patterns from familiar objects and everyday experiences. Although young children may not be able to articulate or describe items and predictable events as patterns, they depend on them to structure their day: "A parent or teacher who changes a pattern gets a quick reaction from a child. Skipping a page in a story, forgetting the treat at the end of a meal, or not following the bedtime ritual to the letter brings a prompt protest" (Copley, 2010, p. 79).

Thinking about patterns is an important early development in mathematical thinking in general and in the later understanding of algebra in particular (Clements, 2004b). When children work with patterns, they learn two important mathematical principles. The first is *stability:* The elements in a pattern stay the same with each repetition (e.g., the elements are always red and blue). The second mathematical principle is that *order* matters: Once a pattern is established, it determines what follows (e.g., blue always comes after red). Children continue exploring these relationships in algebra when they are older.

How Pattern Awareness Develops

During preschool, children develop the ability to recognize, describe, duplicate (copy), extend (add to and fill in), and create their own patterns. Studies show that young children first identify the core unit (the repeating part) in a pattern, which allows them to predict what comes next and fill in or extend it (Klein & Starkey, 2004). Once they master the basic principles (patterns have the same elements in the same order), they can also create their own patterns. Preschoolers begin by working with simple two-element patterns such as ABABAB (blue, red; blue, red; blue, red) or AABBAAB-BAABB (blue, blue, red, red; blue, blue, red, red; blue, blue, red, red):

At outside time on the swings, Xavier chants, "Up, down, up, down, up, down, up, down." His teacher joins in. "That's my swinging pattern," he tells her.

❖

At work time in the construction area, Rachel uses a hammer to alternately pound a white golf tee and a blue golf tee into a piece of Styrofoam.

Later children can recognize, describe, copy, extend, and create somewhat more complex patterns with two elements (AABAABAAB) and three elements (ABCABCABC):[3]

At work time in the art area, Jonah decorates a get-well card for his mother. He divides a row across the top into boxes and fills them in with pastel colors: yellow, pink, lilac; yellow, pink, lilac; yellow, pink, lilac. He repeats this pattern at the bottom.

In addition to working with repeating patterns, preschool children begin to recognize and predict patterns of increases and decreases, especially if adults call attention to them. For example, they realize that as children leave the planning table one by one to begin work time, the number of people at the table decreases (seeing a pattern of decrease). Or, as children get older, they generally get bigger (relating two patterns of increase, age and size). Children are interested in these patterns of increases and decreases because they relate directly to their concrete everyday experiences. A growing awareness of number helps children recognize and describe the pattern in numerical increases or decreases. Later on, the equations they encounter in middle school algebra express relationships between the pattern of changes in one variable with the pattern of changes in another (e.g., increases in "x" correspond to increases or decreases in "y").

---

[3]Mathematicians require at least three repetitions for a pattern to be established since elements may change (expand, drop, rearrange, or reverse) after two repetitions. Therefore, always provide and describe at least three repetitions when introducing a pattern to a child.

Preschoolers begin by working with simple patterns (ABABAB), as this girl does by stringing a pattern of pink and purple beads.

Teaching Strategies That Support an Understanding of Patterns

Understanding how patterns work is basic to many other areas of mathematical exploration in preschool and to the study of algebra at a later age. To support children as they explore repeating patterns and increasing and decreasing patterns, use the teaching strategies described here.

Provide opportunities for children to recognize and describe patterns in the environment

Help children recognize and describe repeating visual patterns they encounter in clothes, quilts, baskets, toys, books, wallpaper, floor tiles, brickwork, and nature. Begin with the simplest two-element patterns (ABABAB or AABBAABB) and gradually introduce more complex patterns with two elements (AABAABAAB) or three elements (ABCABCABC). Have children look for and describe these patterns in the school or on neighborhood walks by playing search games like pattern hunt or pattern I spy:

At small-group time, the children search for patterns in the classroom. Stella points to the striped curtains. Jimmie holds up the napkin basket that has a checkerboard pattern around the outside. Greta finds a diamond pattern at the base of each marker. "The pattern is the same color as the marker," she observes, "so you know what color is inside." Cal points to his shirt. "Look! I'm wearing a pattern!" he says. The other children examine their clothes for patterns too. Marco finds a zigzag pattern on the side of his sneakers, and Fiona shows everyone the checkerboard pattern on the pocket of her shorts. Nelson points to Carla's head and says, "She has a pattern in her hair [band]." Carla runs to the mirror to look. "Yup," she agrees, "Blue and gold and again all around my head."

Be sure to refer to these discoveries as *patterns* and say what makes a pattern a pattern (e.g., "The pattern you found always goes red, yellow; red, yellow; red, yellow"). Help children identify when something is *not* a pattern because the elements are arranged in random order (red, yellow, red, blue, blue, green) or the sequence changes after a couple of repetitions (red, blue; red, blue; red, green):

At small-group time, Leon makes a border around his picture using small squares of construction paper. He alternates two red squares with two black squares. When he is near the end of the last side, he runs out of black squares. He substitutes blue squares but tells his teacher as he points to that section, "It's not a pattern here."

To further help children grasp the concept of pattern, point to the elements and say their order aloud: "The jewels on the king's crown go red, yellow; red, yellow; red, yellow over and over again, around and around his head."[4] Encourage children to point and says the elements together with you. Hearing and saying the words helps call attention to what they are observing, and the rhythm of the repeated sounds matches the visual repetition of the pattern's image. Remember to always identify and describe at least three repetitions of the pattern.

Most children find it easiest to begin with visual patterns (unless they have visual limitations), but there are many patterns that occur and appeal to the other senses. Children should have experiences with all these types of patterns. For example, call attention to patterned sounds in the environment such as the varying pitches in the alarm of an emergency vehicle (low, high; low, high; low, high; and so on). Simple patterns appear in familiar movements, such as bending and thrusting out their legs as children pump on a swing. (In addition to these naturally occurring patterns, see the next strategy for ideas on how to create patterns that appeal to various senses.)

Cycles of recurring events are another type of repeating pattern that children experience in their environment. A clear example is the daily routine, which is a repeating sequence of activities of (relatively) fixed duration. Many programs use activities such as "calendar time" or "today's weather" to introduce cyclical patterns. These are too abstract for preschoolers to understand, and, in fact, the National Research Council (2009) recommends against using calendars since the grid of seven-day weeks confuses children who

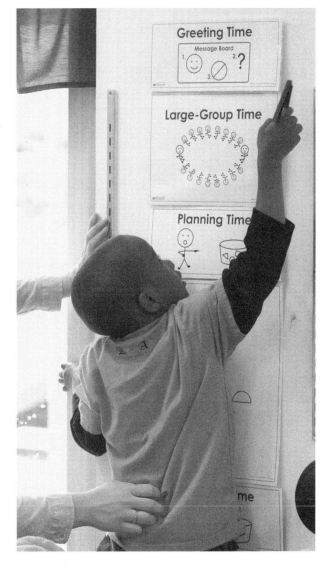

Cycles of recurring events, such as the daily routine, are another type of repeating pattern.

---

[4]Do not recite a pattern by saying "ABABAB" and so on because using letters in this context is confusing for children who are just learning to associate letters with sounds. Instead, read the pattern by naming the actual elements, such as "Blue, red; blue, red; blue, red."

are trying to understand the base-10 system of mathematics. However, young children can begin to make sense of these cycles if they are connected to people and events that are meaningful to them. For example, after regularly using the message board, children will eventually grasp the pattern that at the end of each school week, the two pictures with a drawing of a house represent the weekend, when there is no school. Likewise, children become attuned to the changing weather of the seasons when adults talk with them about the outdoor clothes they are wearing ("It's getting cold outside now that winter is coming, so we'll need to wear our mittens") and/or the types of outdoor play they can engage in (sledding versus swimming):

At outside time, Marta stands in a remaining patch of snow and tells her teacher, "My mommy says spring is almost here and I can help her plant the garden." Marta removes a mitten and wiggles her fingers to show her teacher they'll be free to dig in the dirt when spring comes.

In addition to focusing on repeated patterns, you can also help children recognize and describe increasing and decreasing patterns throughout the day. Again, many of these are inherently of interest to preschoolers. For example, they are fascinated by the idea of getting bigger as they get older. Likewise, they can see decreasing patterns in many parts of the daily routine (e.g., the level of napkins in the basket goes down by one each time the number of napkins on the table increases by one). You can use children's emerging number sense to help them understand these mathematical relationships.

Provide materials and opportunities that lend themselves to creating patterns

Children need materials and activities that encourage them to create patterns based on those they see and those they invent from scratch. Where possible, work with administrators to incorporate patterns in classroom furnishings such as curtains, upholstery, and carpets; the arrangement of light fixtures on the ceiling; or wood inlays in floors and doors. Make sure you provide a variety of patterns in classroom and outdoor materials such as doll and dress-up clothes and costume jewelry; designs on the paper or plastic wrapping around crayons and markers; sample books of wallpaper, carpeting, and brickwork (home furnishing and builder supply stores are often happy to donate outdated design books); and items from nature such as shells, fallen plant leaves and flower petals, tree bark, and photos of wings and fur. Materials that encourage children to make their own patterns include beads, sticks, pattern blocks, small animal or people figures in different sizes and colors, pegs and pegboards, drawing and collage materials, and interactive drawing and pattern-making computer programs. Many books also feature patterns in their cover designs and/or illustrations. Craft magazines and books have pictures of patterns in quilting, weaving, and woodworking projects. Children enjoy figuring out the core patterns in these materials and may be inspired to create their own.

In the following examples, these children create patterns using the everyday materials they find in their classroom environment:

At work time in the toy area, David makes a simple alternating pattern with red and black Connect Four pieces. He says it's like the chessboard his grandpa plays with at home.

❖

At outside time, Lily makes a column of twigs followed by a column of pebbles. She makes six more columns, alternating twigs and pebbles. She collects more when she runs out.

❖

At work time in the house area, Daniel separates small bears into piles of red and blue. He counts to make sure he has 10 in each pile. Then he puts a large basket on the table and alternates throwing a red bear and a blue bear into it. If he misses, he throws the same bear again so as not to break the color order. When all the bears are in the basket, he empties it out, divides the bears into two piles, and starts throwing them into the basket again.

Provide opportunities at small- and large-group times, transitions, and outdoor activities for children to work with simple patterns. For example, at small-group time, offer collage materials they can use to create a border around a picture they have made. At large-group time, preschoolers can create simple movement patterns by alternating where on their bodies they pat to a steady beat or how they move with objects:

At large-group time, Garth waves his streamer up over his head and down around his knees, repeating this sequence several times. His teacher Jason imitates this sequence and, after several repetitions, encourages Garth to come up with words to accompany his movements. "High and low," says Garth. Together they wave their streamers and chant, "High, low, high, low, high, low." "We're moving and singing a pattern," Jason comments to Garth.

Children can also explore sound patterns through alternations in pitch (low and high), tempo (fast and slow), or loudness (loud and soft). Many familiar children's songs have patterns (the chorus repeats after each verse). You can also plant a garden with alternating rows of lettuce and carrots or arrange apple and orange slices in a pattern at snacktime.

Look for opportunities to have fun with patterns

For many children, working with patterns has the same attraction as doing puzzles or constructing with blocks. You can build on this interest by posing fun challenges as you work alongside them. For example, encourage them to fill in the blank — create or copy a pattern such as ABAB__B (using real objects) and ask children what goes in the empty space. Or change an element in the core unit and ask them to change the rest accordingly; for example, change from ABABAB to ACACA__ and ask them to continue the new pattern. A related challenge is to find the error. Deliberately make a mistake in a pattern and see if children correct you. If not, say something like, "This doesn't look right. I hope you can help me fix it."

During large-group time, the children and teachers explore the patterns they can make with rhythm sticks.

Another strategy is to depict a change in the weekly routine on the message board (a three-day weekend) and encourage children to describe how the change contrasts with the regular schedule ("I wonder why there are *three* houses on the board"). You can also call children's attention to increasing or decreasing patterns that emerge in their play; for example, wonder why the level of napkins in the basket is lower for the child who passes out two per place setting than for the child who passes out one. As preschoolers attempt to explain these situations, they construct principles about the properties and functions of patterns — information they need for advanced explorations later on.

For examples of how children work with patterns at different stages of development, and how adults can gently support and extend their learning in this key developmental indicator [KDI], see "Ideas for Scaffolding KDI 38. Patterns" on page 111. The ideas suggested in the chart will help you support and gently extend children's work with patterns as you play and interact with them in other ways throughout the daily routine.

At greeting time, this teacher calls attention to a change in the weekly schedule by drawing three schools with the universal "no" symbol over them. She then asks the children to describe what this means to them.

Ideas for Scaffolding KDI 38. Patterns

Always support children at their current level and occasionally offer a gentle extension.

Earlier	Middle	Later
Children may	*Children may*	*Children may*
• Say something is a pattern when it's not (e.g., point to a string of random colored beads and say, "I made a pattern"). • Line up objects in no particular order; copy a simple pattern.	• Recognize patterns (e.g., point to the stripes on a shirt or the pegs on a pegboard and say, "It makes a pattern!"). • Create or extend a simple pattern (i.e., ABABAB); for example, string beads red, blue; red, blue; red, blue to make a necklace; create a knees, head; knees, head; knees, head sequence for others to copy at large-group time.	• Say *why* something is a pattern (e.g., point to the border they drew on a picture and say, "I made a pattern. It's the same green and red and green and red all around"). • Create or extend a complex pattern (i.e., AABAABAAB or ABCABCABC); for example, Jenna sees that Mitch has made a line of pegs that goes red, blue, green; red, blue, green; red, blue, green and she says to him, "You need a red one next."
To support children's current level, adults can	*To support children's current level, adults can*	*To support children's current level, adults can*
• Label the actual attributes in what children call a pattern (e.g., "The beads go red, blue, green, blue, yellow"). • Line up objects in the same random order as the children.	• Acknowledge when children identify something as a pattern (e.g., "There's a pattern on your shirt"). • Say the pattern elements together with children, repeating them at least three times.	• Encourage children to identify what is and is not a pattern and to say why something is (not) a pattern. • Provide collections of materials for children to create complex patterns (e.g., shells in three or more shapes, buttons in several colors).
To offer a gentle extension, adults can	*To offer a gentle extension, adults can*	*To offer a gentle extension, adults can*
• Look for and label patterns in the environment (e.g., "Look at the pattern in the rug. It goes dot, square; dot, square; dot, square). • Look for opportunities to make and describe patterns (e.g., at work time, say, "I put my dinosaurs in this order: small, big; small, big; small, big").	• Encourage children to find other patterns in the classroom and outdoors (e.g., "Where else do you see a pattern?"). • Model making more complex patterns (e.g., ABCABCABC).	• Make an error in a pattern (e.g., ABCABCABD) to see if children spot it; if not, call their attention to it (e.g., "Oops…something looks wrong"). • Create a different complex pattern and see if children can extend it (e.g., if children make an AABAABAAB pattern, make an AABBAABBAABB pattern).

KDI 39. Data Analysis

E. Mathematics

39. Data analysis: Children use information about quantity to draw conclusions, make decisions, and solve problems.

Description: Children collect, organize, and compare information based on measurable attributes. They represent data in simple ways (e.g., tally marks, stacks of blocks, pictures, lists, charts, graphs). They interpret and apply information in their work and play (e.g., how many cups are needed if two children are absent).

At work time, Ella, Jayla, and Brianna talk with Sue (their teacher) about who will turn five first (all have May birthdays). With Sue's help, Ella writes the three girls' names and the number date of their birthday on a sheet of paper. Ella compares the dates and sees that hers is the lowest. "I'm having my birthday first," she says.

❖

At small-group time, Joey sorts the buttons in his basket by color, counts them, and announces he has more red than any other color. Vinod does the same and determines his blue buttons outnumber the others. "I wonder which color is the most in everyone else's basket," muses Li, their teacher. The children decide to make a chart. They write each child's name and the color of the button they have the most of in their basket, and, with Li's help, analyze the data: Three children have more red, two have more blue, and two have more yellow buttons. "I have more red," says Joey. "I have four, Ian has three, and Maya has three."

Young children engage in data analysis when they describe, organize and compare, represent, interpret, and apply information in their play. With adult support, preschoolers can represent data in simple lists, charts, and graphs, and count and measure to reach conclusions about their own mathematical questions.

The Importance of Data Analysis

In data analysis, young children use their other emerging mathematical skills, for example, in counting and measuring, to solve problems. Classification in science (key developmental indicator [KDI] 46) involves many of the same investigative abilities. However, in science, children may only work with qualitative attributes, such as color or magnetism. In data analysis, the attributes are quantitative; that is, they can be organized and tabulated in some way. Because quantities are involved, data analysis is considered a content area of mathematics.

During the preschool years, data analysis serves the primary purpose of helping young children develop their thinking and reasoning abilities. They learn how to ask questions in ways that can be answered by collecting and tabulating information. Then, once they have gathered the relevant data, children gain practice interpreting and applying the results to problems that interest them. They may be addressing a practical problem, such as checking the sign-in sheet, seeing that two children

> "Data analysis contains one big idea: classifying, organizing, representing, and using information to ask and answer questions."
>
> — Clements (2004b, p. 56)

in their group are absent, and figuring out how many napkins to distribute at snacktime. The classroom community might be seeking information; for example, how big a garden should we dig so everyone who wants can plant a row? Or it may be a matter in which they have an emotional investment; for example how many friends are turning five and going to kindergarten next year instead of coming back to preschool? Asking such questions in quantitative language underlies mathematical reasoning. It is also a skill that applies to many other areas of learning. In fact, opportunities for engaging children in data analysis often arise during other activities, as happened with one group of preschoolers after listening to a favorite book at small-group time:

After reading Caps for Sale *by Esphyr Slobodkina, a group of preschoolers are curious about how many hats they can balance on their heads. They count eight children in the group and six hats in the dress-up area, not enough for each child to have one hat, let alone try to balance several. "How can we solve this problem?" asks the teacher. The children suggest asking parents to donate hats and taking turns with the hats that are there. (They don't think anyone will balance more than six at once!) While each child gives it a try, the others can count and the teacher can help them write down the number on a chart. Then they can reuse the same hats when it is the next child's turn (an important insight for the*

children). Parents are happy to donate old hats, and variations of the cap game become a favorite work-time activity.

How Data Analysis Knowledge and Skills Develop

The development of data analysis skills proceeds in stepwise fashion during the early years and beyond. For example, beginning preschoolers do not automatically use categories to sort and make sense of the information they gather. They will generate a list but not group the data in any way that allows interpretation. Children at this age are only interested in the particulars (Russell, 1991). Later, with their emerging numerical knowledge and intentional guidance from an adult, older preschoolers can begin to sort and count data according to meaningful categories (National Council of Teachers of Mathematics, 2000). For example, here is how children might list what they collected on three nature walks, taken respectively at an earlier, middle, and later time during the school year:

What We Collected on Our Nature Walk		
Developmental Level		
Earlier	*Middle*	*Later*
Sue: Pinecone	Pinecone: Sue, Ada, Zeb	Pinecone: /// (3)
John: Stone	Stone: John, Fiona	Stone: // (2)
Isaac: Leaf	Acorn: Sammy	Acorn: / (1)
Ada: Pinecone	Leaf: Isaac	Leaf: / (1)
Sammy: Acorn		
Fiona: Stone		
Zeb: Pinecone		

Opportunities for engaging children in collecting and analyzing data arise throughout the school day. Preschoolers' natural interest in the properties of objects, with what and whom they play, solving problems with materials, what they eat, how they move, and so on provide openings for framing and answering questions in mathematical terms. For example:

- At message board: How many visitors are coming today?

- At planning time: How many children before me also plan to play in the art area?

- At work time: How many blocks will we need to build a tower exactly as tall as this one?

- At cleanup time: Can we fit more blocks on the shelf if we turn them a different way?

- At recall time: How many children played in the house area today?

- At snacktime: Do children like apple slices more, less, or the same as orange slices?

- At outside time: Who goes fast, faster, and fastest down the slide?

- At small-group time: Are more children using cookie cutters with their play dough or not using them?

- At large-group time: Do four-year-olds take longer steps than three-year-olds?

Data gathering and analysis are also an intriguing way to involve families in mathematical activities. Juanita V. Copley (2010) observes

In this small-group time, the teacher and children are cutting out pictures of different food items they find in the weekly grocery circulars and sorting the pictures into small baskets to find out which type of food (e.g., fruits, vegetables, cereals) they find the most pictures of in the advertisements.

that "once children have collected, organized, and analyzed data from their peers, they often are interested in how their brothers, sisters, parents, or neighbors will answer their questions. So data analysis activities are a natural link between classroom and home" (p. 142). Whether at school or at home, young children can be enthusiastic data gatherers when the questions they ask and the problems they pose are of interest to them.

Teaching Strategies That Support Data Analysis

Manipulating materials is as important in data analysis as it is in other areas of mathematics. Therefore, supplying diverse materials, asking for family contributions, and working with items that children collect provides many bits of data for children to analyze. They do this by sorting and counting objects, recording and describing the results, and applying what they learn to answer questions or solve problems that arise in play. In addition to materials, children can also compare and quantify such things as events, actions, and preferences. To promote children's early explorations in data analysis, use the following teaching strategies.

Provide opportunities to sort and count things and to describe and apply the results

Provide materials that children can group according to the attributes that interest them (e.g., size, color, texture, weight, sound, speed, pattern). As noted earlier in this chapter, classifying objects based on their attributes is also an important activity in science (KDI 46. Classifying). However, in science the emphasis may be qualitative while mathematics involves a quantitative component. For example, a child

classifying might observe that "the shells in this pile have ridges and the stones in that pile are smooth." The conclusion of a child analyzing data might be that "I found more shells than stones on the beach." Both are valuable experiences, but each child is learning and applying a different lesson.

While sorting seems to come naturally to preschoolers, they may not attach a general term (none, some, all) or a specific number value (two, five) to the parts of a collection. Likewise, even after they are able to count, they do not automatically apply or make use of the resulting numbers. Children are unlikely to make sense of quantitative findings unless adults encourage them to think about what the numbers mean and how they can be applied to answering their own questions or solving the problems they encounter in play, as shown in this anecdote:

After the first few snowfalls of the year, several children in Miss Kay's class comment that outside time is getting shorter. When asked why they think this is so, they conclude that they now have to put on snowsuits and boots. "How can we solve this problem?" asks Miss Kay. Rachel suggests they could get ready faster if those who are ready first help the others. "How can we be sure it's faster that way?" Miss Kay wonders. Milo suggests she time them with a stopwatch, and the others like this idea. That day, without children helping one another, Miss Kay times "15 minutes" and writes it down. The next day, when the children help one another, she times and writes "10 minutes." "I know that 10 is smaller than 15," says Ben. The children conclude their solution works and help one another after that.

In this scenario, the teacher asks the class to think about why it is taking longer to go outside after the first snowfall and to problem-solve a way to get ready faster. She then wonders how they will know if their solution works and helps them use mathematics (measuring with a stopwatch) to check out their idea. The children apply their emerging mathematical skills to compare two numbers and conclude that their solution works. They are ready to engage in data analysis, but it would not have happened without the teacher's active support.

Help children represent data using lists, tabulations, charts, and graphs

Recording and tabulating data on lists and simple charts and graphs may seem too abstract for preschoolers. On the contrary, it makes data gathering and analysis tangible for them. Documentation helps children see the results of their calculations in writing (using numerals, checks, or hash marks). Or they can move small objects into columns or baskets and count them. These physical representations and movements make the data analysis process concrete. Without this step, preschoolers' sorting and counting tends to be haphazard. They may enjoy doing it, but it is unlikely to have lasting value or enhance their mathematical thinking and reasoning skills.

Seeing the results in writing makes the outcome of an investigation more significant in a child's eyes. In this respect, children are similar to adults. The chart or graph provides evidence of the effort that went into collecting the data. When adults take time to summarize the results, it further says to children, "The work you did is important!" Summarizing the results also helps them make the connection between their data-gathering efforts and the outcomes they generate.

Children can record data with basic materials such as a sheet of paper on a clipboard, which is also a portable tool for keeping track of things on a classroom hunt or neighborhood walk. A large sheet of paper posted at eye level is useful

for calculating and displaying the results. Using a different color writing tool for each category (brown for stones and green for leaves) can add another visual component. For some children, the act of moving something into one data column or another helps them think through the attributes of what they are sorting. Therefore, small objects (animal figures, blocks, shells) sorted on either side of a line drawn on paper or into small containers can be useful as children collect data on the things that interest them.

Using these materials, young children can record data in several simple ways, with adult help as needed. They can write their names under two or three headings and count the number of names in each list (how many children do and do not like chocolate). Hash marks, numerals, or some other type of representation can be used to indicate the total. A chart divided into two or more columns (or rows or boxes) also works for counting and comparing bits of data (three columns or boxes to compare how many children prefer chocolate, vanilla, or strawberry ice cream). With experience and guidance, many preschoolers can also begin to interpret bar graphs. So, continuing the previous example, they can understand that bars of different heights represent how many children, respectively, prefer each flavor. These same results can also be represented in nonwritten ways. For example, children can move small counters to corners of the table, across dividing line(s) on paper or into baskets to stand for the conditions being compared.

When you start to look for them, you will discover many situations throughout the day that lend themselves to recording and tabulating data. For example, how many children worked in each area, the number of each type of material used in a small-group time construction or collage activity, the color of children's clothing, whether the soles of children's shoes are plain or patterned, the number of people in children's

During recall time, this teacher introduces a chart with the children's names and letter-linked symbols written at the top of each column.

She then shows the children sticky notes labeled with the name and symbol of the different classroom areas.

As each child recalls, the teacher asks the child to find the place (or places) the child worked in on the sticky notes and place the note (or notes) under his or her name.

After everyone has recalled, the teacher and the children look at the chart to see if they can figure out which area had the most (and fewest) children playing in it during work time.

Their analysis of the data: The block area had the most children playing in it during work time!

families, the months in which birthdays occur, and so on. Once they get in the habit of charting things, the children will generate many ideas themselves!

Ask and encourage children to ask questions that can be answered by gathering data

Young children are curious about virtually everything in their experience. While people like to say that preschoolers are always asking "Why?" they can also be encouraged to ask "How many?" This is the type of question that lends itself to gathering data and making sense of the results.

To encourage children to ask data-based questions, begin by asking such questions yourself about things that interest them. For example, if the children in your class enjoy playing with or reading books about cars, you might plan a neighborhood walk and say, "I wonder how many cars we'll see." If children are also interested in trucks, you might pose a question in which children look for both types of vehicles (e.g., "Do you suppose we'll see more cars or more trucks on our walk?"). Take along materials so you and/or the children can record the data, such as one or more clipboards and a writing tool to make a hash mark every time the children see a car (or truck). When

The day after going on a field trip to the local apple orchard, the children explore the attributes of the different types of apples they picked. On a piece of chart paper, the teacher writes down their findings. The children conclude that red apples are juicy and sweet and green apples are sour.

you get back to the classroom, look at the data. Together with the children, count the number of cars. If you are answering the second question, count the number of trucks as well. Help them interpret the results by talking about which type of vehicle they saw more (or fewer) of on the walk, and invite them to offer explanations of why this might be so ("The cars are parked because mommies and daddies are at work. The truck guys are taking stuff to the stores").

Once children understand the idea behind "how many" questions, they will begin to pose such queries themselves. Some might be simple tallies of one item ("How many children wore mittens to school today?" or "How many spoons of white paint do I need to add to my red paint to match this color pink?"). Others will involve making numerical comparison of two or more groups ("How many children wore mittens and how many wore gloves?" or "Does the red paint need more white paint added to match this shade of pink or that shade of pink?").

Sometimes children need help rephrasing an idea or a question so that it can be answered using data analysis. For example, children might wonder who brought a juice box in their lunch and who got milk or whose seed hasn't come up yet, whose has sprouted, and whose has a leaf. You might then say, "So your idea is to count how many children have juice and how many have milk. What if someone doesn't have anything to drink?" In this way, you can frame children's ideas in mathematical terms and help them consider the variables on which they can gather data.

Of course, asking is only the first step in the process of data collection and analysis. You might encourage children to predict the outcome ("Do you think more children drink juice or drink milk?"). Next, work alongside them to gather and tabulate the data (counting juice drinkers and milk drinkers). Finally, children need your guidance to interpret and apply the results ("How many juice boxes and milk cartons should we bring for lunch on our field trip?") This last step gives added reason — in addition to sheer curiosity — for asking the question in the first place. Using the data analysis process to pose questions, collect information, analyze results, and apply the findings helps children appreciate the relevance of mathematics in their own lives.

For examples of how children engage with data analysis at different stages of development, and how adults can scaffold their learning in this KDI, see "Ideas for Scaffolding KDI 39. Data Analysis" on page 122. The ideas suggested in the chart will help you support and gently extend children's use of data analysis as you play and interact with them in other ways throughout the daily routine.

Ideas for Scaffolding KDI 39. Data Analysis

Always support children at their current level and occasionally offer a gentle extension.

Earlier	Middle	Later

Children may

- Contribute ideas or actions unaware of representing data (e.g., randomly put bears on plates to show which areas they played in).
- Group things into two or more sets but not compare their quantities (e.g., make two piles of rocks but not say if one set has the same, more, or fewer/less rocks).
- Look at or listen to data unaware of its meaning (e.g., when another child says, "There are more boys than girls here," say, "I'm here!").

Children may

- Represent data in concrete ways (e.g., put a picture of an apple or a pear in one of two bowls to vote for a favorite fruit).
- Analyze information from sets in general quantitative terms such as none, some, and all or same, more, and fewer/less (e.g., scan the room and observe, "There are more boys than girls here").
- Interpret data (e.g., look at the recall chart and say, "Lots of kids played in the house area today").

Children may

- Represent information (data) in more abstract ways (e.g., write tally marks, checks, or their name on lists, charts, and simple graphs).
- Analyze information from sets in specific quantitative terms (e.g., looking at a list of the pets children have, say, "There are three checks in the dog row but five checks in the cat row").
- Interpret and apply data (e.g., "More kids like apples than pears. Let's buy apples at the market").

To support children's current level, adults can

- Accept what children say or do (e.g., "You put lots of bears on the plates").
- Provide materials that can be grouped into collections; group things alongside children.
- Acknowledge what children share ("Yes, you are here!").

To support children's current level, adults can

- Provide opportunities to represent data in concrete ways (e.g., at a transition, say, "Those with sneakers gather by the blocks; those with sandals at the sink").
- Repeat none, some, and all (or more, same, fewer/less) to describe sets (e.g., "More like vanilla than chocolate").
- Affirm interpretations of data (e.g., "Yes, the house area was the most popular today").

To support children's current level, adults can

- Provide materials (e.g., clipboards, chart paper) and opportunities (e.g., signing up for snack choices) to represent data.
- Acknowledge specific quantitative terms to analyze data (e.g., "You figured out how many children played in each area").
- Repeat children's data conclusions (e.g., "You're right. Blue is the favorite color, so we'll need more blue paper").

To offer a gentle extension, adults can

- Model representing data (e.g., "I played with you in the toy area, so my bear goes on the plate with the toy area symbol").
- Describe and encourage children to describe sets in general quantitative terms (e.g., "There are more beads in this pile than that pile").
- Label what data represent (e.g., "The checks show this many played on the swing").

To offer a gentle extension, adults can

- Model abstract ways to represent data (e.g., "You ride the bus, so your name goes under the bus picture").
- Encourage children to analyze sets using specific quantities (e.g., "I wonder how many want milk and how many want juice?").
- Encourage children to apply lessons from data (e.g., "If more like pretzels than raisins, what should we put in the trail mix?").

To offer a gentle extension, adults can

- Encourage children to represent data (e.g., "How should we show the orders for fries and salad?").
- Analyze more complex data (e.g., charts with three or four columns).
- Encourage children to interpret and apply more complex data (e.g., "Lots of children played with bikes and balls, but no one played with hoops. What toys should we get out tomorrow?").

Mathematics: A Summary

General teaching strategies for mathematics

- Provide a wide variety of mathematics materials in every area of the classroom.
- Converse with children using mathematics words and terms.
- Encourage children to use mathematics to answer their own questions and solve their own problems.
- Pose challenges that encourage mathematical thinking.

Teaching strategies that support using number words and symbols

- Use numeral words to describe everyday materials and events.
- Call attention to numerals (number symbols) in the environment.
- Encourage children to write numerals.

Teaching strategies that support counting

- Count and compare everything.
- Provide materials to explore one-to-one correspondence.
- Engage children in simple numerical problem solving.

Teaching strategies that support understanding part-whole relationships

- Provide materials that can be grouped and regrouped.
- Provide materials that can be taken apart and put back together.

Teaching strategies that support naming and using shapes

- Provide shapes for children to see and touch.
- Encourage children to create and transform shapes and observe and describe the results.
- Name shapes and the actions children use to transform them.

Teaching strategies that support spatial awareness

- Provide materials and plan activities that encourage children to create spaces.
- Encourage children to handle, move, and view things from different perspectives.
- Use and encourage children to use words that describe position, direction, and distance.

Teaching strategies that support measuring

- Support children's interest in identifying and comparing measurable attributes.
- Encourage children to estimate quantities.
- Use and encourage children to use measurement words.

Teaching strategies that support an understanding of unit

- Support children's use of conventional and unconventional measuring tools.
- Model accurate measuring techniques.

Teaching strategies that support an understanding of patterns

- Provide opportunities for children to recognize and describe patterns in the environment.
- Provide materials and opportunities that lend themselves to creating patterns.
- Look for opportunities to have fun with patterns.

Teaching strategies that support data analysis

- Provide opportunities to sort and count things and to describe and apply the results.
- Help children represent data using lists, tabulations, charts, and graphs.
- Ask and encourage children to ask questions that can be answered by gathering data.

This small-group time gives children an opportunity to measure (with an eyedropper) and count (how many drops to put in).

References

Baroody, A. J. (2000). Does mathematics instruction for three- to five-year olds really make sense? *Young Children, 55*(4), 61–67.

Baroody, A. J. (2004). The developmental bases for early childhood number and operations standards. In D. H. Clements, J. Samara, & A.-M. DiBiase (Eds.), *Engaging young children in mathematics: Standards for early childhood mathematics education* (pp. 173–219). Mahwah, NJ: Lawrence Erlbaum.

Baroody, A. J., Lai, M.-L., & Mix, K. S. (2006). The development of young children's early number and operation sense and its implications for early childhood education. In B. Spodek & O. Saracho (Eds.), *Handbook of research on the education of young children* (pp. 187–221). Mahwah, NJ: Lawrence Erlbaum.

Benigno, J. P., & Ellis, S. (2004). Two is greater than three: Effects of older siblings on parental support of preschoolers' counting in middle-income families. *Early Childhood Research Quarterly, 19*(1), 4–20.

Bowerman, M. (1996). Learning how to structure space for language: A cross-linguistic perspective. In P. Bloom, M. A. Peterson, L. Nadel, & M. F. Garrett (Eds.), *Language and space* (pp. 385–436). Cambridge, MA: MIT Press.

Campbell, P. F. (1997). Connecting instructional practice to student thinking. *Teaching Children Mathematics, 4,* 106–110.

Campbell, P. F. (1999). Fostering each child's understanding of mathematics. In C. Seefeldt (Ed.), *The early childhood curriculum: Current findings in theory and practice* (3rd ed., pp. 106–132). New York, NY: Teachers College Press.

Chalufour, I., & Worth, K. (2003). *Discovering nature with young children*. St. Paul, MN: Redleaf Press.

Clark, C. A., Pritchard, V. E., & Woodward, L. J. (2010). Preschool executive functioning abilities predict early mathematics achievement. *Developmental Psychology, 46*(5), 1176–1191. doi:10.1037/a0019672

Clements, D. H. (1999). The effective use of computers with young children. In J. V. Copley (Ed.), *Mathematics in the early years* (pp. 119–128). Reston, VA: National Council of Teachers of Mathematics and Washington, DC: National Association for the Education of Young Children.

Clements, D. H. (2004a). Geometric and spatial thinking in early childhood education. In D. H. Clements, J. Sarama, & A.-M. DiBiase (Eds.), *Engaging young children in mathematics: Standards for early childhood mathematics education* (pp. 267–297). Mahwah, NJ: Lawrence Erlbaum.

Clements, D. H. (2004b). Major themes and recommendations. In D. H. Clements, J. Sarama, & A.-M. DiBiase (Eds.), *Engaging young children in mathematics: Standards for early childhood mathematics education* (pp. 7–72). Mahwah, NJ: Lawrence Erlbaum.

Clements, D. H., & Sarama, J. (2007). Early childhood mathematics learning. In F. K. Lester, Jr. (Ed.), *Second handbook of research on mathematics teaching and learning* (pp. 461–555). New York, NY: Information Age.

Clements, D. H., Sarama, J., & DiBiase, A.-M. (Eds.). (2004). *Engaging young children in mathematics: Standards for early childhood mathematics education*. Mahwah, NJ: Lawrence Erlbaum.

Clements, D. H., & Stephan, M. (2004). Measurement in pre-K to grade 2 mathematics. In D. H. Clements, J. Sarama, & A.-M. DiBiase (Eds.), *Engaging young children in mathematics: Standards for early childhood mathematics education* (pp. 299–317). Mahwah, NJ: Lawrence Erlbaum.

Clements, D. H., Swaminathan, S., Hannibal, M. A. Z., & Sarama, J. (1999). Young children's concepts of shape. *Journal for Research in Mathematics Education, 30,* 192–212.

Copley, J. V. (2010). *The young child and mathematics* (2nd ed.). Washington, DC: National Association for the Education of Young Children and Reston, VA: National Council for Teachers of Mathematics.

Copple, C. (2004). Mathematics curriculum in the early childhood context. In D. H. Clements, J. Sarama, & A.-M. DiBiase (Eds.), *Engaging young children in mathematics: Standards for early childhood mathematics education* (pp. 83–87). Mahwah, NJ: Lawrence Erlbaum.

Duncan, G. J., Dowsett, C. J., Claessens, A., Magnuson, K., Huston, A. C., Klebanov, P., …Brooks-Gunn, J. (2007). School readiness and later achievement. *Developmental Psychology, 43*(6), 1428–1446. doi:10.1037/0012-1649.43.6.1428

Epstein, A. S. (2009). *Numbers Plus Preschool Mathematics Curriculum*. Ypsilanti, MI: HighScope Press.

Epstein, A. S. (2012a). *Language, literacy, and communication*. Ypsilanti, MI: HighScope Press.

Epstein, A. S. (2012b). *Science and technology*. Ypsilanti, MI: HighScope Press.

Epstein, A. S., & Hohmann, M. (2012). *The HighScope Preschool Curriculum*. Ypsilanti, MI: HighScope Press.

Fuson, K. C. (1988). *Children's counting and concept of number*. New York, NY: Springer-Verlag.

Gardner, H. (1991). *The unschooled mind: How children think and how schools should teach*. New York, NY: Basic Books.

Gelman, R., & Gallistel, C. R. (1986). *The child's understanding of number*. Cambridge, MA: Harvard University Press. (Original work published in 1978)

Gersten, R., & Chard, R. (1999). Number sense: Rethinking arithmetic instruction for students with mathematical disabilities. *Journal of Special Education, 33*(1), 18–28. doi:10.1177/002246699903300102

Ginsburg, H. P., Greenes, C., & Balfanz, R. (2003). *Big math for little kids: Prekindergarten and kindergarten*. Parsippany, NJ: Dale Seymour.

Ginsburg, H. P., Inoue, N., & Seo, K-H. (1999). Young children doing mathematics: Observations of everyday activities. In J. V. Copley (Ed.), *Mathematics in the early years* (pp. 88–99). Reston, VA: National Council of Teachers of Mathematics and Washington, DC: National Association for the Education of Young Children.

Greenes, C. (1999). Ready to learn: Developing young children's mathematical powers. In J. Copley (Ed.), *Mathematics in the early years* (pp. 39–47). Reston, VA: National Council of Teachers of Mathematics and Washington, DC: National Association for the Education of Young Children.

Greenes, C., Ginsburg, H. P., & Balfanz, R. (2004). Big math for little kids. *Early Childhood Research Quarterly, 19*(1), 159–166. doi:10.1016/j.ecresq.2004.01.010

Jones, S. S., & Smith, L. B. (2002). How children know the relevant properties for generalizing object names. *Developmental Science, 2*, 219–232. doi:10.1111/1467-7687.00224

Kamii, C. (2000). *Young children reinvent arithmetic* (2nd ed.). New York, NY: Teachers College Press.

Kirova, A., & Bhargava, A. (2002). Learning to guide preschool children's mathematical understanding: A teacher's professional growth. *Early Childhood Research and Practice, 4*(1), 1–21.

Klein, A., & Starkey, P. (2004). Fostering preschool children's mathematical knowledge: Findings from the Berkeley Math Readiness Project. In D. H. Clements, J. Sarama, & A.-M. DiBiase (Eds.), *Engaging young children in mathematics: Standards for early childhood mathematics education* (pp. 343–360). Mahwah, NJ: Lawrence Erlbaum.

Klibanoff, R. S., Levine, S. C., Huttenlocher, J., Vasilyeva, M., & Hedges, L. V. (2006). Preschool children's mathematical knowledge: The effect of teacher "math talk." *Developmental Psychology, 42*(1), 59–69.

Lehrer, R. (2003). Developing understanding of measurement. In J. Kilpatrick, W. G. Martin, & D. Schifter (Eds.), *A research companion to principles and standards for school mathematics* (pp. 179–192). Reston, VA: National Council of Teachers of Mathematics.

Levine, S. C., Suriyakham, L. W., Rowe, M. L., Huttenlocher, J., & Gunderson, E. A. (2010). What counts in the development of young children's number knowledge? *Developmental Psychology, 46*(5), 1309–1319.

Miller, K. F., Smith, C. M., Zhu, J., & Zhang, H. (1995). Preschool origins of cross-national differences in mathematical competence: The role of number naming systems. *Psychological Science, 6*, 56–60. doi:10.1111/j.1467-9280.1995.tb00305.x

Mix, K. S. (2002). The construction of number concepts. *Cognitive Development, 17*, 1345–1363. doi:10.1016/S0885-2014(02)00123-5

Mix, K. S., Levine, S. C., & Huttenlocher, J. (1999). Early fraction calculation ability. *Developmental Psychology, 35*, 164–174. doi:10.1037/0012-1649.35.1.164

National Association for the Education of Young Children & National Council of Teachers of Mathematics. (2002, April). *Early childhood mathematics: Promoting good beginnings. A joint position statement of NAEYC and NCTM*. Washington, DC: National Association for the Education of Young Children.

National Council of Teachers of Mathematics. (2000). *Principles and standards for school mathematics*. Reston, VA: Author.

National Council of Teachers of Mathematics. (2006). *Curriculum focal points for prekindergarten through grade eight mathematics*. Reston, VA: Author.

National Mathematics Advisory Panel. (2008). *Foundations for success: The final report of the National Mathematics Advisory Panel.* Washington, DC: US Department of Education.

National Research Council. (2009). *Mathematics learning in early childhood: Paths toward excellence and equity.* Washington, DC: National Academies Press.

Neill, P. (n.d.). *Meaningful math in preschool: Making math count throughout the day.* Unpublished manuscript.

Russell, S. J. (1991). Counting noses and scary things: Children construct their ideas about data. In D. Vere-Jones (Ed.), *Proceedings of the third international conference on teaching statistics* (Vol. 1, pp. 158–164). Voorburg, Netherlands: International Statistical Institute.

Sarama, J., & Clements, D. H. (2009). *Early childhood mathematics education research: Learning trajectories for young children.* New York, NY: Routledge.

Seo, K.-H. (2003). What children's play tells us about teaching mathematics. *Young Children, 58*(1), 28–34.

Seo, K.-H., & Ginsburg, H. P. (2004). What is appropriate in early childhood mathematics education? In D. H. Clements, J. Sarama, & A.-M. DiBiase (Eds.), *Engaging young children in mathematics: Standards for early childhood mathematics education* (pp. 91–104). Mahwah, NJ: Lawrence Erlbaum.

Sierpinska, A. (1998). Three epistemologies, three views of communication: Constructivism, social cultural approaches, interactionism. In H. Steinbring, M. G. Bartonlini Bussi, & A. Sierpinska (Eds.), *Language and communication in the mathematics classroom* (pp. 30–62). Hillsdale, NJ: Erlbaum.

Sophian, C. (2004). Mathematics for the future: Developing a Head Start curriculum to support mathematics learning. *Early Childhood Research Quarterly, 19*(1), 59–81. doi:10.1016/j.ecresq.2004.01.015

Sophian, C., Wood, A. M., & Vong, K. I. (1995). Making numbers count: The early development of numerical inferences. *Developmental Psychology, 31*(2), 263–273. doi:10.1037//0012-1649.31.2.263

Tomlinson, H. B., & Hyson, M. (2009). Developmentally appropriate practice in the preschool years — ages 3–5: An overview. In C. Copple & S. Bredekamp (Eds.), *Developmentally appropriate practice in early childhood programs serving children from birth through age 8* (3rd ed., pp. 111–148). Washington, DC: National Association for the Education of Young Children.

Uttal, D. H., & Wellman, H. M. (1989). Young children's representation of spatial information acquired from maps. *Developmental Psychology, 25,* 128–138.

Van de Walle, J. A. (1998). *Elementary and middle school mathematics: Teaching developmentally* (3rd ed.). New York, NY: Addison Wesley Longman.

Zur, O., & Gelman, R. (2004). Young children can add and subtract by predicting and checking. *Early Childhood Research Quarterly, 19*(1), 121–137. doi:10.1016/j.ecresq.2004.01.003